Resource Management in Agroecosystems

Edited by Gabrijel Ondrasek and Ling Zhang

Published in London, United Kingdom

Resource Management in Agroecosystems
http://dx.doi.org/10.5772/intechopen.104280
Edited by Gabrijel Ondrasek and Ling Zhang

Contributors
Melekote Nagabhushan Arun, Rapolu Mahender Kumar, Banugu Sreedevi, Guntupalli Padmavathi, Pallakonda Revathi, Dayyala Srinivas, Boya Venkatesh, Neha Pathak, Shakkie Kativu, Clemence Zimudzi, Evelyn Ngarakana, Brita Stedje, Nasir Shad, Ling Zhang, Zohra Nasheen, Rabia Afza, Amir Kassam, Rachid Mrabet, Akashdeep Singh, Tarun Sharma, Theodor Friedrich, Gottlieb Basch, Rachid Moussadek, Emilio Gonzalez-Sanchez, Tirivashe Phillip Masere, Addisu Fufa Ebbisa

First published in London, United Kingdom, 2023 by IntechOpen
IntechOpen is the global imprint of INTECHOPEN LIMITED, registered in England and Wales, registration number: 11086078, 5 Princes Gate Court, London, SW7 2QJ, United Kingdom

British Library Cataloguing-in-Publication Data
A catalogue record for this book is available from the British Library

Additional hard and PDF copies can be obtained from orders@intechopen.com

Resource Management in Agroecosystems
Edited by Gabrijel Ondrasek and Ling Zhang
p. cm.
Print ISBN 978-1-83768-420-5
Online ISBN 978-1-83768-421-2
eBook (PDF) ISBN 978-1-83768-422-9

For EU product safety concerns:
IN TECH d.o.o., Prolaz Marije Krucifikse Kozulić 3, 51000 Rijeka, Croatia,
info@intechopen.com or visit our website at intechopen.com.

Meet the editors

Gabrijel Ondrasek is a full professor with more than 20 years of experience at the Faculty of Agriculture, University of Zagreb, where he has served as the head of the Department of Soil Amelioration for more than 10 years. Currently, he coordinates international master's studies at the university and participates in numerous councils, committees, and boards. With a research focus on sustainable soil and water management in agroecosystems, salinization, and metal contamination processes, Dr. Ondrasek has spent more than 35 months at prestigious research and academic institutions in Australia, Europe, and Latin America. He has edited 4 scientific books and 8 special issues in scientific journals and published 18 book chapters and more than 100 peer-reviewed articles.

Dr. Ling Zhang obtained his Ph.D. in Soil Science from Nanjing Agricultural University, China, and visited the Department of Ecology and Evolutionary Biology, Rice University, USA, from July 2012 to December 2013. Dr. Zhang is now a professor at the College of Forestry, Jiangxi Agricultural University, China, supervising graduate students. Dr. Zhang studies global change biology, forest ecology, plant invasion, and soil carbon and nitrogen cycling. He has published more than eighty papers and authored or co-authored several books related to global change ecology or soil ecology in recent years. Dr. Zhang is an associate editor and reviewer of several journals.

Contents

Preface

In an era marked by rising global populations, climate changes, and mounting environmental concerns, the sustainable management of our Earth's resources has become an imperative. In this context, agriculture plays a pivotal role as a significant user of resources and a primary source of human food supply. The intricate interplay between agriculture and the environment has spurred a growing need for resource management strategies that balance the demands of food production with the preservation of (agro)ecosystems.

Agroecosystems, the dynamic amalgamation of agriculture and the surrounding environment, serve as the primary arena for this exploration. The ever-evolving challenge of feeding a global population of more than nine billion people by 2050 while safeguarding the planet's ecological integrity necessitates innovative approaches. This book sheds light on the complex issues surrounding resource management within agroecosystems and serves as a comprehensive guide for researchers, policymakers, practitioners, and students alike.

The journey through these pages will traverse the multifaceted terrain of resource management, offering insights into sustainable agricultural practices, more efficient water use, soil conservation, energy optimization, and biodiversity preservation. It will scrutinize the intricate relationships between land, water, energy, and biological resources, showcasing the interdependencies that govern agroecosystems. Throughout, the book will draw upon the latest research findings, case studies, and real-world examples to underscore the significance of resource management in achieving food security, economic stability, and environmental stewardship.

Our intention is to not only provide a thorough understanding of the challenges we face but also to spotlight the opportunities that arise from adopting resource-efficient and ecologically sound practices. As we navigate the pages ahead, we will encounter the voices of experts, practitioners, and visionaries who have dedicated their efforts to redefining the boundaries of sustainable agriculture. Their experiences, perspectives, and innovations will inspire us all to engage with the urgent task of resource management in agroecosystems.

We extend our heartfelt gratitude to all the authors who have shared their knowledge and experiences, the reviewers who provided valuable feedback, and the publishers who have made this book possible. As we embark on this intellectual journey, we hope that *Resource Management in Agroecosystems* serves as a compass guiding us toward a more sustainable and resilient future. May it inspire readers to become active participants in the vital endeavor of balancing the needs of humanity with the well-being of our planet.

Gabrijel Ondrasek
Faculty of Agriculture,
University of Zagreb,
Zagreb, Croatia

Ling Zhang
Key Laboratory of Silviculture,
College of Forestry,
Jiangxi Agricultural University,
Nanchang, China

Chapter 1

Conservation Agriculture: Climate Proof and Nature Positive Approach

Rachid Mrabet, Akashdeep Singh, Tarun Sharma, Amir Kassam, Theodor Friedrich, Gottlieb Basch, Rachid Moussadek and Emilio Gonzalez-Sanchez

Abstract

The development pathways of countries and regions have impacted land-climate interactions and shaped challenges, opportunities and actions. Adverse impacts of climate change increasingly threaten livelihoods and resilience of people around the globe, food security and the stability of environmental resources. Globally, the current food systems are not fit for purpose. Land-based options such as Conservation Agriculture (CA) were found to mitigate climate change, regenerate soils and ensure durable food systems. Achieving sustained results using CA systems, under climate change and social pressures, while maximizing co-benefits related to food and nutrient security, social and biological diversity, ecosystem restoration and services and sustainable development, requires appropriate country-specific policies and significant investment. CA implementation is challenging and context specific and necessitates an integrated framework and road map to enable deeper ambitions for social equity and development and inclusive economic growth.

Keywords: no-till, soil mulch cover, climate change, sustainability, environment, carbon sequestration

1. Introduction

The interaction between land and climate is a complex system thoroughly influencing the agriculture production systems around the globe [1]. Agricultural production systems are the largest single source of environmental degradation, responsible between 21 and 37% of global greenhouse gas (GHG) emissions through deforestation, depletion of soil carbon, release of nitrous oxide and enteric fermentation. Without intervention, these are likely to increase by about 30–40% by 2050, due to increasing demand based on population and income growth and dietary change. Agriculture is also responsible for 70% of freshwater use, 30% of energy use and

80% of land conversion [2]. Conventional agricultural practices revolved around the burning of crop residues to facilitate land preparation for the succeeding crop, regular plowing and tillage of the land for preparing seedbeds and controlling weeds. Reduced natural soil productivity and pest control were corrected with new high yielding breeds, fertilizers and pesticides. These practices initially had a positive effect on production and yield of crops but at the cost of continuous land degradation, erosion by wind or water, underground water pollution, oxidation of the soil organic matter due to tillage and emitting carbon dioxide (CO2) in large amounts [3]. Like in other aspects of the economy a trickle-down of benefits to poor farmers is assumed, but rarely materialized.

By the year 2050, the global population is expected to increase to 9.1 billion which would mean that the existing production systems need to gear up and increase their food production by 70 per cent by the year 2050, assuming food waste and change of consumer preferences continue unchanged. Producing sufficient food with finite resources to feed the growing global population while having a smaller impact on the environment has always been a great challenge. In addition, the 2022 IPCC reports generated enormous attention as a demand for immediate actions across all sectors and regions. There is a need for rethinking the actual food systems and address all the connected challenges and threats and explore the root causes of unsustainability. Consequently, healthy growth and stable productivity of crops and livestock require innovative models of food production for resource-saving, environmentally friendly agriculture. Conservation Agriculture (CA) has proven to overcome the shortcomings of tillage-based agriculture in terms of sustainability as a promising system-based approach [4]. Here, we review the environmental impacts of CA that should lead to a paradigm shift in goals and models of food production for promoting sustainable and regenerative agriculture worldwide.

2. Conservation agriculture: adoption evolution and trends

The need for a transformation of conventional tillage-based agriculture became obvious in the early 1930s after the 'Dust Bowl' trembled the mid-west farming communities of the United States and obliged the scientific community to reorient its research agenda and focus more on erosion mitigation and soil conservation through no-tillage systems (later called Conservation Agriculture or CA systems) [5]. CA is a resource-conserving agricultural concept that is steadily gaining ground and covers an estimated area of 205 million hectares (14.7% of global cropland) (also see **Table 1** for regional distribution). This represents an increase of 93% in global CA cropland area since 2008/09 and represents an annual increase of about 10 Mha.

The major countries practicing CA in 2018/19 are the USA (44.0 Mha), Brazil (43.9 Mha), Argentina (32.9 Mha), Australia (22.9 Mha), Canada (21.7 Mha) and others (39.6 Mha) [5]. In other terms, the total CA area is approaching 70% and 75% of the total cropland area in South America and in Australia, respectively. However, since 2008/09, percentage change in CA adoption has been greater in Asia, Africa and Europe than in the other continents, and corresponds to 33.1 Mha or about16% of the global CA cropland area [5]. CA as climate proof agriculture and its roles for soil sustainability and resilience are widely recognized and should favor increase in its adoption by mainstreaming the concept in agricultural and environmental policies.

DOI: http://dx.doi.org/10.5772/intechopen.108890

Region	CA cropland area 2008/2009	CA cropland area 2013/2014	CA cropland area 2015/2016	CA cropland area 2018/2019	Percent change in CA area since 2015/2016	Percent change in CA area since 2013/2014	Percent change in CA area since 2008/2009	Percent CA cropland area in the region 2018/2019
S and C America	49,564.10	66,377.00	69,895.00	82,996.18	18.7	25.0	67.5	68.7
North America	40,003.80	53,967.00	63,181.00	65,937.22	4.4	22.2	64.8	33.6
Australia and New Zealand	12,162.00	17,857.00	22,665.00	23,293.00	2.8	30.4	91.5	74.0
Russia and Ukraine	100.00	5200.00	5700.00	6900.00	21.1	32.7	6800.0	4.5
Europe	1560.10	2075.97	3558.20	5601.53	57.4	169.8	259.0	5.2
Asia	2630.00	10,288.65	13,930.20	17,529.02	25.8	70.4	566.5	3.6
Africa	485.23	993.44	1509.24	3143.09	108.3	216.4	547.8	1.1
Total	106,505.23	156,759.06	180,438.64	205,400.04	13.8	31.0	92.9	14.7

Table 1.
Global spread of CA cropland area ('000 ha) in different regions for 2008/2009, 2014/2015, and 2018/2019, and corresponding percent change (source: [5]).

3. CA as climate proof agriculture

Since food production is sensitive to weather conditions, the very existence of mankind is being threatened by an unseen force referred to as climate change. Climate change is expected to adversely affect the climatic/weather phenomena thereby impacting the global food supply system [6, 7]. The evidence is irrefutable that GHG are choking our planet and placing billions of people in danger. An increasing number of people are not able to realize their right to adequate food. In 2020, between 720 and 811 million people in the world faced hunger, up to 161 million more than in 2019 [8]. In other words, climate change, food security and biodiversity are the "trilemma of land use". Solutions to these challenges should be integrated to combine and tackle multiple goals. Hence, considering the world is at stake, various international organizations such as Food and Agriculture Organization (FAO), World Bank, and many more, have come together to tackle climate change and food insecurity and search for a reasonable, economical, and sustainable solution. In addition, these urgencies should also rely on the need for grassroots led structural change to stay within the ecological boundaries of the planet, several of which have already been exceeded.

Ingenious and meritoriously employed land-based measures [9], including specific measures to protect and enhance soil organic carbon stocks, can directly support the global environmental and sustainability goals under the UNFCCC [1], the UNCCD [10], and the CBD [11]. In lieu, among these measures, CA works as a systemic approach with its key contributions to sustainability, climate change adaptation and mitigation as well as food security [12–15].

From its wide-ranging adaptation and adoption, CA systems are being practiced in rainfed and irrigated systems, annual, perennial and mixed cropping systems, orchards and plantation systems, agroforestry systems, pasture systems, organic and nonorganic systems, and rice-based systems [16, 17].

Abundant literature and multi-stakeholder innovation platforms across various farming systems showed that CA is a climate proofing agriculture [18–20]. In fact, several recent studies have found that fully-implemented CA can improve crop yield stability—a measure of climate resilience—in different soil types, climates and cropping systems. Worldwide, CA has helped bolstering productivity, augmenting resilience to weather shocks, and tumbling negative externalities (i.e. [21] in USA; [22] in Australia; [23–25] in India; [26, 27] in China; [28–31] in Africa; [32, 33] in West Asia and North Africa, [34] in Europe). As sometimes observed, decrease in crop yield following the adoption of CA largely depends on whether CA has been correctly implemented, with the use of appropriate seeders, seed rates, fertilizer applications and management practices followed to manage weeds and pests. Some yield reductions in initial years were also due to problems of drainage and stagnation of water in cool and humid regions as result of poorly structured soils from a tillage-based farming history.

Experiences in drought conditions have shown that CA yields can be twice as much as conventional agriculture, peaking up to 4-fold higher yields in wheat [19, 35]. Sun et al. [36] found that in arid regions, CA permitted both increased carbon sequestration and crop yields. Based upon a meta-analysis comprising 610 studies, 48 crops and 63 countries, Pittelkow et al. [37] found variable responses from CA compared to conventional tillage systems. The authors concluded that CA are better performing under a range of crop species in arid regions – particularly where water is limiting to crop growth. The authors also reported that yield gaps are due to partial use of CA principles, which obviously will not produce all the CA benefits. When no-till is combined with residue retention and crop rotation, which is the full implementation of the CA principles, no significant yield reduction is noticed: indeed, this combination of techniques significantly increases crop yields in dry climates. A dataset containing 4403 paired (CA vs. CT) yield observations collected between 1980 and 2017 for eight major staple crops in 50 countries presented by Su et al. [38] also confirmed this trend. In addition, selecting high-efficiency crop varieties and optimizing agronomic (nitrogen) management practices to increase water/nitrogen use efficiency is an effective way to increase crop yield with less associated environmental costs under CA. In order to achieve increased yield stability across climate and soil gradients, it is of paramount importance to grow mixtures of crop species or mixtures of genotypes to exploit positive interaction effects and thus reduce the risk of crop failure [39].

A meta–analysis using data from 9686 paired site–year comparisons across South Asia in a variety of cropping systems found that, CA systems provided 5.8% higher mean yield than conventional agricultural practices [40]. In another study by Laik et al. [41], under CA systems, yields of wheat and rice increased by 46–54 and 10–24%, respectively, over conventional tillage, thereby obtaining ~53% higher total output from the CA system. In a review study by Das et al. [24], the CA systems increased yields of crops from 2% to 200% depending on crop rotations and years of implementation.

Through a meta-analysis of 933 observations from 16 different countries in sub-Saharan African studies, Corbeels et al. [42] showed that average yields under CA are only slightly higher than those of conventional tillage systems (3.7% for six major crop species and 4.0% for maize). Larger yield responses for maize result from mulching and crop rotations/intercropping. They also concluded that when CA principles are implemented concomitantly, maize yield increases by 8.4%, which proves the fact, that the lower yield benefits reported in the study resulted from mixing CA systems with systems that only adopted some of the CA principles.

One of the most entrenched benefits of CA systems is their ability to improve soil water storage. The maintenance of crop residues and mulches at the surface of soils under CA systems improves the water balance of the soil-cropping system. CA systems improve the uptake, conservation, and use of available water in the soil by the crops [43, 44]. All this increases the responsiveness of CA systems to changes in climate, meaning crops under these systems have a much better capacity for coping and adapting to drought. Under rainfed ecologies of eastern and south African countries, CA systems reduced the yield variability by 11% over CT [45].

When CA systems are implemented in warmer and drier regions, higher crop yields are often observed due to a lowering of soil temperatures in addition to increases in soil water storage. In irrigated regions, higher water storage and better water management under CA systems can reduce the amount of water required for crop production and help conserve water resources [46–48].

4. Environmental sustainability: soil re-carbonization, conservation, health and security

Soil's multi-functionality and health were generally neglected to address food and climate security challenges [3, 49]. However, after the Paris Agreement was signed, stakeholders committed in a voluntary action plan to implement farming systems and practices that maintain or enhance soil carbon stocks in agricultural soils and to preserve carbon-rich soils [50]. Global technical potential of SOC sequestration is 1.45–3.44 Pg C/year (2.45 Pg C/year) but varies with type of soils, management and ecologies [51].

The push to the CA-based system is due to its environmental and productivity sustainability and especially its ability to (i) reduce soil degradation, erosion and runoff, (ii) mitigate greenhouse gas emissions and (iii) sequester atmospheric CO_2 in the form of soil organic carbon, tackle climate change, (iv) improve biodiversity below and above the soil surface, and (v) enhance production system resilience to abiotic and biotic stresses [52–57]. In fact, CA systems were initially adopted for soil conservation and erosion control benefits, but they are gaining more and more attention as a practice to maintain and/or increase SOC and harness ecosystem services in agroecosystems [58].

CA aims to implement soil-based strategies and long-term soil fertility dynamics that restore soil functions and health [3, 56, 59] and increase carbon storage reversing consequently the food insecurity spiral [30, 60–62]. In other terms, it is beneficial for crop production and soil health and functions and hence to global food security and adaptation of agriculture to climate change [16, 17, 38, 62–65]. Lal et al. [66] concluded that evidence-based strategy based on CA can allow re-carbonization of depleted soils. Studies by Blanco-Canqui and Ruis [67] confirmed that when CA systems are applied in an integrative way, synergic effects of the principles give rise to levels of soil organic matters.

It goes without saying that carbon stored in the soil is the most stable carbon (C) pool, an essential part of ecosystem services and a tool to tackle climate change [68]. In view of its role in soil aggregation and erosion control, in availability of plant nutrients and in ameliorating other forms of soil degradation than erosion, CA systems have proven to reduce soil degradation and rebuild soil quality. However, in areas with low fertility, integrated nutrient management is essential to ensure a build-up of SOC and the success of CA systems (i.e., in Africa). This cycle can be broken by judicious addition of nutrients to the soil/crop system via organic or synthetic fertilizers and/or the incorporation of legumes into cropping rotations [56, 69].

In lieu of climate change, sequestering CO_2 has become inevitable. CA systems in comparison to the conventional practices saw an increase in SOC in top-soil (0–15 cm) by 3.8 Mg ha^{-1}, in the deepest layer (70–100 cm) by 2.5 Mg ha^{-1} and mean C sequestration rates of 0.09 and 0.27 Mg ha^{-1} yr^{-1} [70]. Soil carbon sequestration bids to improve soil fertility and reduce carbon dioxide levels in the atmosphere. Among continents, Africa is the smallest contributor to greenhouse gas emissions but is highly susceptible to climate change, which is mainly responsible for rising temperatures, fluctuating rainfall patterns, increased frequency of disastrous events such as droughts and floods leading to heavy losses in terms of resources. Gonzalez-Sanchez et al. [71] reported that an estimate of the potential annual carbon sequestration in African agricultural soils through CA amounts to 143 Tg of C per year, that is 524 Tg of CO_2 per year. This figure represents about 93 times the current sequestration figures. In addition, this potential is almost 3 times higher than the one found for Europe by Gonzalez-Sanchez et al. [72], which amounts to 189 Tg CO_2 per year.

In the rice-wheat cropping system, an improvement in carbon stocks by 20% and 40% at a depth of 0–15 and 15–30 cm was realized by following the CA principles [73]. A worldwide meta-analysis by Li et al. [64] found that, on average, the number of water stable aggregates in CA systems are 31% greater compared to conventionally tilled systems. Such soil quality improvements are based on greater SOM content which provides greater abundance of habitats to support microbial, micro- and meso-fauna activity. By enhancing soil health and re-carbonizing the soils [3], CA systems establish dynamic ecological conditions in the soil/plant/landscape continuum which offers resilient performance with maximum productivity (water and nutrient use efficiency and water productivity) [4].

Several authors reported that CA systems minimize on-site and off-site effects with regards to soil degradation and that benefits to soil health and ecosystems follow a chain-like process. Under CA systems, erosion is lessened, infiltration is improved, and water losses either through evaporation or runoff are reduced, allowing the crop to have more water in dry periods or years [64]. In other terms, CA also contributes to the environment by mitigating pollution as it reduces off-site transport of residual agrochemicals through runoff and soil sediments. This reduces the surface transport of nitrate and phosphorus from agricultural fields and the eutrophication of water bodies. Also leaching of nutrients under CA is usually reduced, as the water is mainly transported through macro pores (bypass flow) and not washing the soil matrix as long as synthetic or organic fertilizer or slurry is not applied directly before a heavy rainstorm, which can potentially increase leaching of nitrate to groundwater through the macropores [58, 74].

According to Lal [74], in addition to carbon sequestration and erosion control, adoption of CA systems accentuates several other ecosystem services such as biodiversity, elemental cycling, and resilience to natural and anthropogenic perturbations, all of which can affect food security. It was also reported that CA systems do not lead to significant compaction and higher bulk densities than traditional systems based on soil disturbance [64].

In addition, when combined with frontier technologies (precision agriculture, plant breeding and biotechnology, microbial biotechnology, smart fertilizers, biochar additions etc.), CA systems can help to soak up even more carbon in the soil, create soil resilience to achieve food security and mitigate climate change and allow higher and stable yields [54, 75, 76].

5. Economics under CA systems: no regret options

Countries seek to and should improve the well-being of people and especially farmers. The conventional system of agricultural production is hugely dependent on intensive tillage operations with the support of much labor or heavy farm machinery. The latter results in higher $CO2$ emissions and both in higher production costs [77]. Reducing the tillage operations has the potential of reducing emissions and fuel consumption. CA systems can save up to 80% of fossil fuel energy used by tillage [24].

The farmers, and mainly the resource-poor ones, need production systems that are regenerative, reliable, financially viable and profitable. However, many scientific studies agree that CA systems are cost-effective, energy efficient and allow farmers higher and more stable incomes [24, 34]. The major factor leading to lower costs in CA systems is attributed to bypassing soil manipulation and disturbance unlike conventional tillage systems, where 4–5 primary and secondary tillage operations are performed for seedbed preparation and weed control, which acquire higher costs [78, 79].

Even if CA systems in the beginning might have undesirable effects on crop yield levels, the cost of cultivating crops decreases with fewer use of machinery and compensates for eventual initial yield declines. Subsequently, continuous use of such practices improves soil properties, sustains crop productivity and ultimately economic returns [80]. In fact, according to several authors, there is mounting evidence that when CA is inconsistently applied, it leads to lower yields and higher costs than expected [38, 62, 81–83]. Arenas-Calle et al. [84] showed that the lack of climate-smartness resulted in yield penalties in early stages of CA implementation. However, in eastern and southern Africa the highest financial returns (90–95%) from CA investments by small–holder farmers were realized under low-rainfall conditions (<700 mm), thereby providing clear evidence of the climate smartness of CA systems under soil moisture–stressed conditions [45].

With the reduced expenses in terms of labor, energy and monetary inputs, CA practices reduce the cost of cultivation. Reduced expenditure in such a pattern was observed in winter wheat for no-tillage practices (1300 Yuan ha^{-1}), reduced tillage (2250 Yuan ha^{-1}) as compared to conventional tillage practices (2500 Yuan ha^{-1}) [85]. Especially in the case of small and resource poor farmers, with reduced usage of machinery cost (<65.52%) under CA, farmers spend less (14.46%) on different cultivation practices, increasing their net returns as compared to conventional agriculture practices [86, 87]. In sub-Saharan Africa, scientific studies revealed that with systematic use of practices such as no-tillage, residue retention and crop rotation the costs of cultivating maize or soybean were reduced (20–29%) and the net returns, the benefit-cost ratio increased to a greater extent [88–98]. A similar impact of less soil disturbing practices such as permanent beds and zero tillage was obtained on net returns or profitability of maize-chickpea rotation in India (28.8% and 24% respectively) [99–103].

In a regional study in Ethiopia, CA was found to have reduced the labor usage by 32–41% whereas 50–60% labor was replaced at the critical periods of crop production due to reduced tillage operations in the maize-soybean intercropping system. Further, a maximum return of 15,545 ETH birr ha^{-1} and 12,693 ETH birr ha^{-1} was obtained when soybean and haricot were intercropped in maize [104]. The net returns in production of the rice CA systems were 581 USD ha^{-1} in comparison to 412 USD ha^{-1} under the conventional system. The gross returns in the rice-wheat system were highest (2456 USD ha^{-1}) under the CA system [79].

Choudhary et al. [90] found 22.3 and 24.5% higher grain yield of pearl millet [*Cenchrus americanus* (L.) Morrine] and Indian mustard [*Brassica juncea* (L.) Czernj.] under CA systems, respectively, compared to conventional systems, which ultimately led to higher net returns (US$ 1270 ha^{-1}).

From a meta-analysis carried out by Ogle et al. [98], it was concluded that CA systems drastically reduce the number of field operating hours and associated fuel use by about 69%. From these studies, it is clear that broadening access to finance, including international and climate finance will catalyze adoption and accelerate the shift towards CA systems.

6. Conclusion

Research and development efforts in agriculture have been increasingly oriented towards improving modern, industrial or corporate agriculture—new chemicals, hybrid and genetically modified seeds, mechanization, factory farming, etc. Hence, the agriculture sector is replete with innovations but not all of them were found sustainable.

The CA systems backed by various institutions, research scholars, policymakers were found able to adapt to the fast-changing environment thus making the food system healthy, flexible, productive and profitable. Further, CA helps to extenuate the greenhouse gas emissions and increase the carbon stocks making soils resilient, reliable and sustainable. In other words, the main benefits of CA systems cover numerous areas and contribute to a number of SDGs. In addition, CA feasibility or adoption was assessed in contrasted biophysical, social and economic environments.

CA systems are alternative pathways for agriculture to be more conducive to durable food systems and longer-term sustainability. Especially soil carbon sequestration and health improvement allowed by CA systems can support various ecosystem services related to climate change adaptation, food security and biodiversity due to enhanced soil fertility and nutrient pools, increased moisture retention, improved water availability to plants and reduced soil erosion and runoff [58]. The number of countries explicitly including SOC in agricultural land (including wetlands) in the Nationally Determined Contributions (NDCs) increased from 28 (15% of first-round NDCs assessed) to 35 (24% of latest NDCs assessed) [101] which is still insufficient. An international agenda for restoring soil health and inclusion of soil carbon sequestration in policies and actions should be advocated and supported [88, 102]. Policies promoting the target of land degradation neutrality can support food security, human wellbeing and climate change adaptation and mitigation [2].

Barriers to the adoption of CA system are more related to farmers' attributes (adopter's characteristics, limited availability of resources, level of perception, mind-set, cultural values, illiteracy, willingness for change, etc.) and their enabling environment (e.g., legal compliance, governance, lack of training and capacity building, stakeholder communication, lack of financial support, insufficient economic and social incentives) than to technical concerns (i.e. herbicide and machinery availability and costs, energy use and price, competitive uses of crop residues and livestock etc.). The science related to CA systems is currently advanced enough to inform the formulation of policy and incentive programs for CA adoption at a scale large enough to result in the radical transformation of mainstream agricultural production systems CA [5, 103].

Dis-adoption, accumulating challenges and difficulties of mainstreaming CA by additional farmers arise from two main issues: (i) CA is dynamic, meaning that it should respond to simultaneous changes in environment, social and/or economic contexts, (ii) CA is also a holistic concept based on a system-wide approach to solving farm management shifts and problems while considering the integrality of the food system. In addition, agriculture functions are changing over time and getting more complex with increasing socio-economic and environmental stresses and social and institutional shocks [33]. Approaches for upscaling CA range from sophisticated decision support systems to improved enabling environments (i.e., through land policies and subsidies focused on water, environment, and poverty) and promotion of social or sustainability-oriented learning processes [16].

The new Green Revolution (GR) of the twenty-first century must be: (i) soil-centric, based on soil health and resilience, (ii) ecosystem-centric, based on eco-efficiency of inputs, (iii) knowledge or innovation-centric, based on scientific principles, and (iv) nature-centric, based on nature positive solutions which restore and enhance nature [95]. The new GR should also recognize the "One Health" concept, which states that the "health of soil, plants, animals, people, ecosystems, and the planetary processes is one and indivisible [94].

The 8th World Congress on Conservation Agriculture (WCCA), which inspired from these paradigms of the new GR, set a goal to increase the global CA cropland area to 50% of the total cropland by 2050, in particular to respond to the global challenge to mitigate the advancing climate change and land degradation and reduce gaps in food security and nutrition (as well as other sustainable development goals). This represents an area of 700 M ha [92, 103]. In achieving such goals, policy and economic incentives should be enforced and augmented in most countries. In addition, the integration of CA benefits in the farming system (e.g., value chain design, marketing, labeling), can lead to giving carbon both economic and environmental values and thus increasing farmer income and stewardship. Social norms as well as psychological and behavioral factors must be considered for widespread adoption of CA systems. Accordingly, a multi-stakeholder engagement and joint coordination (i.e., science-policy dialog and engaged civil society) are major issues in the development and implementation of a CA Road Map for wide mainstreaming and large-scale adoption by farmers in markedly diverse ecologies. Implementing CA Road maps enable governments, landowners and land managers, and the community to share responsibility for land-based challenges mitigation and hence in achieving or reaching SDGs. According to Lal [93], sustainable intensification of agroecosystems (which includes CSA systems) can produce enough food grains to feed one person for a year on 0.045 ha of arable land. Hence, another issue of prime importance concerns socializing CA for the small land size farmers while integrating livestock and trees mainly in Africa and Asia. There is great momentum in merging principles of CA with those regenerative types of farming and especially those related to tillage, synthetic fertilizers and pesticide use [97]. However, issues related to GMOs are still largely debated within the agroecological stream. Kassam and Kassam [4] proposed an inclusive ethical and responsible system to integrate CA systems with plant-based diets and organic farming practices in order to move from corporate agriculture.

Author details

Rachid Mrabet[1], Akashdeep Singh[2], Tarun Sharma[2], Amir Kassam[3*], Theodor Friedrich[4], Gottlieb Basch[5], Rachid Moussadek[6] and Emilio Gonzalez-Sanchez[7,8,9]

1 National Institute of Agricultural Research, INRA, Rabat, Morocco

2 Department of Agronomy, CSK Himachal Pradesh Agricultural University, Palampur, India

3 University of Reading, UK

4 Former Residential Representative, FAO, Bolivia

5 University of Evora, Portugal

6 International Center for Agricultural Research in Dry Areas (ICARDA), Rabat, Morocco

7 Universidad de Córdoba, Spain

8 European Conservation Agriculture Federation (ECAF), Brussels, Belgium

9 Asociación Española Agricultura de Conservación Suelos Vivos (AEACSV), Spain

*Address all correspondence to: amirkassam786@googlemail.com

References

[1] Jia G, Shevliakova E, Artaxo P, De Noblet-Ducoudré N, Houghton R, House J, et al. Land–climate interactions. In: Climate Change and Land: An IPCC Special Report on Climate Change, Desertification, Land Degradation, Sustainable Land Management, Food Security, and Greenhouse Gas Fluxes in Terrestrial Ecosystems. Geneva, Switzerland: IPCC; 2019

[2] IPCC. Climate Change and Land: An IPCC Special Report on Climate Change, Desertification, Land Degradation, Sustainable Land Management, Food Security, and Greenhouse Gas Fluxes in Terrestrial Ecosystems. Geneva, Switzerland: IPCC; 2019

[3] Montgomery DR. Soil health and the revolutionary potential of Conservation Agriculture. In: Kassam A, Kassam L, editors. Rethinking Food and Agriculture: New Ways Forward. Elsevier; 2020. pp. 219-229

[4] Kassam A, Kassam L. Rethinking Food and Agriculture: New Ways Forward. Switzerland: Elsevier; 2020

[5] Kassam A, Friedrich T, Derpsh R. State of the global adoption and spread of conservation agriculture. In: Kassam A, editor. Advances in Conservation Agriculture, Adoption and Spread. Cambridge, UK: Burleigh Dodds; 2022

[6] Aryal JP, Sapkota TB, Khurana R, Khatri-Chhetri A, Rahut DB, Jat ML. Climate change and agriculture in South Asia: Adaptation options in smallholder production systems. Environment, Development and Sustainability. 2019;**2019**:1-31

[7] Myers SS, Zanobetti A, Kloog I. Increasing CO2 threatens human nutrition. Nature. 2014;**510**:139-142

[8] FAO. The State of the World's Land and Water Resources for Food and Agriculture – Systems at Breaking Point. Rome: FAO; 2021

[9] Roe S, Streck C, Beach R, Busch J, Chapman M, et al. Land-based measures to mitigate climate change: Potential and feasibility by country. Global Change Biology. 2021;**27**:6025-6058. DOI: 10.1111/gcb.15873

[10] Chotte J, Aynekulu E, Cowie A, Campbell E, Vlek P, Lal R, et al. Realising the Carbon Benefits of Sustainable Land Management Practices: Guidelines for Estimation of Soil Organic Carbon in the Context of Land Degradation Neutrality Planning and Monitoring. Bonn, Germany: United Nations Convention to Combat Desertification (UNCCD); 2019

[11] IPBES. Summary for Policymakers of the Assessment Report on Land Degradation and Restoration of the Intergovernmental Science Policy Platform on Biodiversity and Ecosystem Services. Bonn, Germany: IPBES; 2018

[12] Barasa PM, Botai CM, Botai JO, Mabhaudhi T. A review of climate-smart agriculture research and applications in Africa. Agronomy. 2021;**11**:1255. DOI: 10.3390/agronomy11061255

[13] Corbeels M, Naudin K, Guibert H, Torquebiau E, Cardinael R. Is the 4 per thousand initiative for soil organic carbon storage achievable in sub-Saharan Africa? Insights from agroforestry and conservation agriculture. Soil Till Research. 2018;**188**:16-26. DOI: 10.1016/j.still.2018.02.015

[14] Mbow C, Rosenzweig C, Barioni LG, et al. Food security. In: Shukla PR, Skea J, Calvo Buendia E,

Masson-Delmotte V, Pörtner H-O, et al., editors. Climate Change and Land: An IPCC Special Report on Climate Change, Desertification, Land Degradation, Sustainable Land Management, Food Security, and Greenhouse Gas Fluxes in Terrestrial Ecosystems. Geneva, Switzerland: IPCC; 2019

[15] Neufeldt H, Jahn M, Campbell BM, Beddington JR, DeClerck F, De Pinto A. Beyond climate-smart agriculture: Toward safe operating spaces for global food systems. Agriculture & Food Security. 2013;**2**(1):12

[16] Kassam A, editor. Advances in Conservation Agriculture: Systems and Science. Cambridge, UK: Burleigh Dodds; 2020a

[17] Kassam A, editor. Advances in Conservation Agriculture: Practice and Benefits. Cambridge, UK: Burleigh Dodds; 2020b

[18] Kassam AH, Mkomwa S, Friedrich T. Weather-proofing agriculture with conservation agriculture. In: Kassam A, Mkomwa S, Friedrich T, editors. Conservation Agriculture for Africa: Building Resilient Farming Systems in a Changing Climate. Wallingford, UK: CAB International; 2017

[19] Michler JD, Baylis K, Arends-Kuenning M, Mazvimavi K. Conservation agriculture and climate resilience. Journal of Environmental Economics and Management. 2019;**93**:148-169

[20] Ngachan V, Das A. Climate proofing of agriculture for Food security. In: Das et al, editor. Conservation Agriculture for Advancing Food Security in Changing Climate. New Delhi, India: Today & Tomorrow's Printers and Publishers; 2018. pp. 37-57

[21] Williams A, Jordan NR, Smith RG, Hunter MC, Kammerer M. A regionally-adapted implementation of conservation agriculture delivers rapid improvements to soil properties associated with crop yield stability. Scientific Reports. 2018;**8**:8467

[22] Bellotti B, Rochecouste JF. The development of conservation agriculture in Australia—Farmers as innovators. International Soil Water Conservation Research. 2014;**2**:21-34. DOI: 10.1016/S2095-6339(15)30011-3

[23] DasTK,SaharawatYS,BhattacharyyaR, Sudhishri S, Bandyopadhyay K, Sharma AR, et al. Conservation agriculture effects on crop and water productivity, profitability and soil organic carbon accumulation under a maize–wheat cropping system in the North–western Indo–Gangetic Plains. Field Crops Research. 2018;**215**:222-231

[24] Das TK, Ghosh S, Das A, Sen S, Datta D, Ghosh S, et al. Conservation agriculture impacts on productivity, resource-use efficiency and environmental sustainability: A holistic review. Indian Journal of Agronomy. 2021;**66**:111-127

[25] Pradhan A, Chan C, Roul PK, Halbrendt J, Sipes B. Potential of conservation agriculture (CA) for climate change adaptation and food security under rainfed uplands of India: A transdisciplinary approach. Agricultural Systems. 2018;**163**:27-35. DOI: 10.1016/j.agsy.2017.01.002

[26] Du Z, Angers DA, Ren T, Zhang Q, Li G. The effect of no-till on organic C storage in Chinese soils should not be overemphasized: A meta-analysis. Agriculture, Ecosystems and Environment. 2017;**236**:1-11

[27] Lal R. Sustainable intensification of China's agroecosystems by conservation

agriculture. International Soil Water Conservation Research. 2018b;**6**:1-12. DOI: 10.1016/j.iswcr.2017.11.001

[28] Corbeels M, de Graaff J, Ndah TH, Penot E, Baudron F, Naudin K, et al. Understanding the impact and adoption of conservation agriculture in Africa: A multi-scale analysis. Agricultural Ecosystem Environment. 2014;**187**:155-170. DOI: 10.1016/j.agee.2013.10.011

[29] Kuyah S, Sileshi GW, Nkurunziza L, Chirinda N, Ndayisaba PC, Dimobe K, et al. Innovative agronomic practices for sustainable intensification in sub-Saharan Africa: A review. Agronomy for Sustainable Development. 2021;**41**:16. DOI: 10.1007/s13593-021-00673-4

[30] Mkomwa S, Kassam A, editors. Conservation Agriculture in Africa: Climate Smart Agricultural Development. Wallingford, UK: CAB International; 2021

[31] Mupangwa W, Mutenje M, Thierfelder C, Nyagumbo I. Are conservation agriculture (CA) systems productive and profitable options for smallholder farmers in different agro-ecoregions of Zimbabwe? Renewable Agriculture and Food Systems. 2016;**32**(1):87-103

[32] Devkota M, Singh Y, Yigezu YA, Bashour I, Mussadek R, Mrabet R. Conservation agriculture in the drylands of the Middle East and North Africa (MENA) region: Past trend, current opportunities, challenges, and future outlook. Advances in Agronomy. 2021;**172**. DOI: 10.1016/bs.agron.2021.11.001

[33] Mrabet R, Moussadek R, Devkota M, Lal R. No-tillage farming in Maghreb region: Enhancing agricultural productivity and sequestrating carbon in soils. In: Lal R, editor. Advances in Soil Science: Soil Organic Matter and Feeding the Future: Environmental and Agronomic Impacts. United States: CRC Press Taylor and Francis Group; 2021

[34] Gonzalez-Sanchez EJ, Kassam A, Basch G, Streit B, Holgado-Cabrera A, Trivino-Tarradas P. Conservation agriculture and its contribution to the achievement of Agri-environmental and economic challenges in Europe. Aims Agricultural Food. 2016;**1**:387-408. DOI: 10.3934/agrfood.2016.4.387

[35] Mrabet R. No-tillage agriculture in West Asia & North Africa. In: Tow PG, Cooper IM, Partridge I, Birch CJ, editors. Rainfed Farming Systems. Dordrecht Netherlands: Springer; 2011. pp. 1015-1042

[36] Sun WJ, Canadell JG, Yu LJ, Yu LF, Zhang W, Smith P, et al. Climate drives global soil carbon sequestration and crop yield changes under conservation agriculture. Global Change Biology. 2020;**26**(6):3325-3335

[37] Pittelkow CM, Linquist BA, Lundy ME, Liang X, van Groenigen KJ, Lee J, et al. When does no-till yield more? A global meta-analysis. Field Crops Research. 2015a;**183**:156-168

[38] Su Y, Gabrielle B, Makowski D. The impact of climate change on the productivity of conservation agriculture. Nature Climate Change. 2021a;**11**:628-633. DOI: 10.1038/s41558-021-01075-w

[39] Knapp S, van der Heijden MGA. A global meta-analysis of yield stability in organic and conservation agriculture. Nature Communications. 2018;**9**:3632

[40] Jat HS, Choudhary KM, Nandal DP, Yadav AK, Poonia T, Singh Y, et al. Conservation agriculture–based sustainable intensification of cereal systems leads to energy conservation, higher productivity

and farm profitability. Environmental Management. 2020;**65**(6):774-786

[41] Laik R, Sharma S, Idris M, Singh AK, Singh SS, Bhatt BP, et al. Integration of conservation agriculture with best management practices for improving system performance of the rice–wheat rotation in the Eastern Indo–Gangetic Plains of India. Agriculture, Ecosystems and Environment. 2014;**195**:68-82

[42] Corbeels M, Naudin K, Whitbread AM, Kuhne R, Letourmy P. Limits of conservation agriculture to overcome low crop yields in sub-Saharan Africa. Nature Food. 2020;**1**:447-454

[43] Basch G, Kassam A, Friedrich T, Santos FL, Gubiani PI, Calegari A, et al. Sustainable soil water management systems. In: Lal R, Stewart BA, editors. Soil Water and Agronomic Productivity, Advances in Soil Science. Boca Raton, FL: CRC Press; 2012. pp. 229-289

[44] Gonzalez-Sanchez EJ, Veroz-Gonzalez O, Moreno-Garcia M, Gomez-ArizacMR,Ordoñez-FernandezR, Trivino-Tarradas P, et al. Climate change adaptability and mitigation with conservation agriculture. In: Kassam A, Kassam L, editors. Rethinking Food and Agriculture: New Ways Forward. Switzerland: Elsevier; 2020. pp. 231-246

[45] Nyagumbo I, Mupangwa W, Chipindu L, Rusinamhodzi L, Craufurd P. A regional synthesis of seven–year maize yield responses to conservation agriculture technologies in Eastern and Southern Africa. Agriculture, Ecosystems and Environment. 2020;**295**:106898

[46] Basch G, Saavedra N. Poupança de água na cultura do milho através da cobertura do solo. Revista Agrotec. 2022;**42**:16-19

[47] Mrabet R, Moussadek R, Fadlaou A, van Ranst E. Conservation agriculture in dry areas of Morocco. Field Crops Research. 2012;**132**:84-94

[48] Page KL, Dang YP, Dalal RC, Reeves S, Thomas G, Wang W. Changes in soil water storage with no-tillage and crop residue retention on a vertisol: Impact on productivity and profitability over a 50-year period. Soil Tillage Research. 2019;**194**:104319

[49] Shahane AA, Shivay YS. Soil health and its improvement through novel agronomic and innovative approaches. Frontier in Agronomy. 2021;**3**:680456. DOI: 10.3389/fagro.2021.680456

[50] Lal R. Beyond COP 21: Potential and challenges of the "4 per thousand" initiative. Journal of Soil Water Conservation. 2016a;**71**:20A-25A

[51] Lal R. Digging deeper: A holistic perspective of factors affecting SOC sequestration. Global Change Biology. 2018a;**24**:3285-3301. DOI: 10.1111/gcb.14054

[52] Chatterjee S, Chakraborty R, Chakraborty H. Economic impact assessment of conservation agriculture on small and marginal farm households in eastern India. Agricultural Economics Research Review. 2020a;**33**:127-138

[53] Day S, Santos A, Cremonesi M, Maia L, Bartz MLC. Biodiversity management practices and benefits in conservation agriculture systems. In: Kassam A, editor. Advances in Conservation Agriculture. Vol. 2. Cambridge, UK: Burleigh Dodds; 2020. pp. 267-302

[54] Kassam A. Conservation agriculture for regenerative and resilient production systems. In: Uphoff N, Thies JE, editors. Biological Approaches to Regenerative Soil Systems. United States: CRC Press; 2022

[55] Page KL, Dang YP, Menzies NW, Dalal RC. No-till Systems to sequester soil carbon: Potential and reality. In: Dang YP, Dalal RC, Menzies NW, editors. No-till Farming Systems for Sustainable Agriculture: Challenges and Opportunities. Cham, Switzerland: Springer; 2020a. pp. 301-317

[56] Page KL, Dang YP, Dalal RC. The ability of conservation agriculture to conserve soil organic carbon and the subsequent impact on soil physical, chemical, and biological properties and yield. Frontiers in Sustainable Food System. 2020b;**4**:31

[57] Valkama E, Kunypiyaeva Zhapayev R, Karabayev M, Zhusupbekov E, Perego A, Schillaci C, et al. Can conservation agriculture increase soil carbon sequestration? A modelling approach. Geoderma. 2020;**369**:114298

[58] Jayaraman S, Dang YP, Naorem A, Page KL, Dalal RC. Conservation agriculture as a system to enhance ecosystem services. Agriculture. 2021;**11**:718. DOI: 10.3390/agriculture11080718

[59] Somasundaram J, Reeves S, Wang W, Heenan M, Dalal RC. Impact of 47 years of no tillage and stubble retention on soil aggregation and carbon distribution in a vertisol. Land Degradation and Development. 2017;**28**:1589-1602

[60] Jat HS, Datta A, Choudhary M, Sharma PC, Yadav AK, Choudhary V, et al. Climate smart agriculture practices improve soil organic carbon pools, biological properties and crop productivity in cereal-based systems of North-West India. Catena. 2019;**181**:104059. DOI: 10.1016/j.catena.2019.05.005

[61] Richards M, Sapkota T, Stirling C, Thierfelder C, Verhulst N, Friedrich T, et al. Conservation Agriculture: Implementation Guidance for Policymakers and Investors. Rome: FAO; 2014

[62] Su Y, Gabrielle B, Beillouin D, Makowski D. High probability of yield gain through conservation agriculture in dry regions for major staple crops. Scientific Reports. 2021b;**11**:3344. DOI: 10.1038/s41598-021-82375-1

[63] Demenois J, Torquebiau E, Arnoult MH, Eglin T, Masse D, Assouma MH, et al. Barriers and strategies to boost soil carbon sequestration in agriculture. Frontiers in Sustainable Food System. 2020;**4**:37

[64] Li Y, Li Z, Cui S, Jagadamma S, Zhang QP. Residue retention and minimum tillage improve physical environment of the soil in croplands: A global meta-analysis. Soil and Tillage Research. 2019;**194**:104292

[65] Lipper L, Thornton P, Campbell BM, Baedeker T, Braimoh A, Bwalya M, et al. Climate-smart agriculture for food security. Nature Climate Change. 2014;**4**:1068-1072

[66] Lal R, Monger C, Nave L, Smith P. The role of soil in regulation of climate. Philosophical Transactions on Royal Society London B Biological Science. 2021;**376**(1838):20210420

[67] Blanco-Canqui H, Ruis SJ. No-tillage and soil physical environment. Geoderma. 2018;**326**:164-200. DOI: 10.1016/j.geoderma.2018.03.011

[68] Navarro-Pedreño J, Almendro-Candel MB, Zorpas AA. The increase of soil organic matter reduces global warming, myth or reality? Science. 2021;**3**(18):18. DOI: 10.3390/sci3010018

[69] Veloso MG, Angers DA, Tiecher T, Giacomini S, Dieckow J, Bayer C. High

carbon storage in a previously degraded subtropical soil under no-tillage with legume cover crops. Agriculture, Ecosystems and Environment. 2018;**268**:15-23. DOI: 10.1016/j.agee.2018.08.024

[70] Krauss M, Wiesmeier M, Don A, Cuperus F, Gattinger A, Gruber S, et al. Reduced tillage in organic farming affects soil organic carbon stocks in temperate Europe. Soil and Tillage Research. 2022;**216**:105262. DOI: 10.1016/j.still.2021.105262

[71] Gonzalez-Sanchez EJ, Veroz-Gonzalez O, Conway G, Moreno-Garcia M, Kassam A, Mkomwa S, et al. Meta-analysis on carbon sequestration through conservation agriculture in Africa. Soil & Tillage Research. 2019;**190**:22-30

[72] González-Sánchez EJ, Moreno-García M, Kassam A, Holgado-Cabrera A, Triviño-Tarradas P, Carbonell Bojollo R, et al. Conservation agriculture: Making climate change mitigation and adaptation real in Europe. In: European Conservation Agriculture Federation (ECAF). Brussels, Belgium; 2017

[73] Dey A, Dwivedi BS, Bhattacharyya R, Datta SP, Meena MC, Jat RK, et al. Effect of conservation agriculture on soil organic and inorganic carbon sequestration and lability: A study from a rice-wheat cropping system on a calcareous soil of the eastern Indo-Gangetic Plains. Soil Use and Management. 2020;**36**:429-438

[74] Lal R. Enhancing ecosystem services with no-till. Renewable Agricultural Food System. 2013;**28**:102-114

[75] Paustian K, Lehmann J, Ogle S, Reay D, Robertson GP, Smith P. Climate-smart soils. Nature. 2016;**532**:49-57

[76] Paustian K, Larson E, Kent J, Marx E, Swan A. Soil C sequestration as a biological negative emission strategy. Frontiers in Climate. 2019;**1**:8

[77] Voltr V, Wollnerová J, Fuksa P, Hruška M. Influence of tillage on the production inputs, outputs, soil compaction and GHG emissions. Agriculture. 2021;**11**:456

[78] El Gharras O, El Mourid M, Boulal H. Conservation agriculture in North Africa: Experiences, achievements and challenges. In: Kassam A, Mkomwa S, Friedrich T, editors. Conservation Agriculture for Africa: Building Resilient Farming Systems in a Changing Climate. Wallingford, UK: CAB International; 2017. pp. 127-138

[79] Jat RK, Sapkota TB, Singh RG, Jat ML, Kumar M, Gupta RK. Seven-year-old conservation agriculture in rice-wheat rotation of eastern Gangetic Plains of South Asia: Yield trends and economic profitability. Field Crops Research. 2014;**164**:199-210. DOI: 10.1016/j.fcr.2014.04.015

[80] Fowler R, Rockstrom J. Conservation tillage for sustainable agriculture an agrarian revolution gathers momentum in Africa. Soil & Tillage Research. 2001;**61**:93-107

[81] Brown B, Nuberg I, Llewellyn R. Stepwise frameworks for understanding the utilisation of conservation agriculture in Africa. Agricultural Systems. 2017;**153**:11-22

[82] Rosenstock TS, Lamanna C, Chesterman S, Bell P, Arslan A, Richards M, et al. The Scientific Basis of Climate-smart Agriculture: A Systematic Review Protocol. Copenhagen, Denmark: CGIAR; 2016

[83] Yigezu YA, El-Shater T, Boughlala M, Devkota M, Mrabet R, Moussadek R. Can

an incremental approach be a better option in the dissemination of conservation agriculture? Some socioeconomic justifications from the drylands of Morocco. Soil & Tillage Research. 2021;**212**:105067

[84] Arenas-Calle LN, Whitfield S, Challinor AJ. A climate smartness index (CSI) based on greenhouse gas intensity and water productivity: Application to irrigated Rice. Frontier in Sustainable Food System. 2019;**3**:1-13

[85] Su Z, Zhnag J, Wu W, Cai D, Lv J, Jiang G, et al. Effects of conservation tillage practices on winter wheat water-use efficiency and crop yield on the Loess Plateau, China. Agricultural Water Management. 2007;**87**:307-314

[86] Chatterjee S, Ghosh S, Pal P. Soil Carbon Restoration through Conservation Agriculture. London, UK: IntechOpen; 2020b

[87] Khan R, Biswas S, Kundu CK, Jana K. Effect of conservation tillage on yield and economics of fodder crops. International Journal of Environment and Climate Change. 2020;**10**(12):529-539

[88] Amelung W, Bossio D, de Vries W, Kögel-Knabner I, Lehmann J, Amundson R, et al. Towards a global-scale soil climate mitigation strategy. Nature Communications. 2020;**11**:5427. DOI: 10.1038/s41467-020-18887-7

[89] Bhan S, Behera UK. Conservation agriculture in India-problems, prospects and policy issues. International Soil and Water Conservation Research. 2014;**2**(4):1-12

[90] Choudhary M, Rana KS, Bana RS, Ghasal PC, Choudhary GL, Jakhar P, et al. Energy budgeting and carbon footprint of pearl millet–mustard cropping system under conventional and conservation agriculture in rainfed semi–arid agro–ecosystem. Energy. 2017;**141**:1052-1058

[91] FAO, IFAD, UNICEF, WFP, WHO. The State of Food Security and Nutrition in the World 2020. Italy: Rome, FAO; 2020

[92] Kassam A, Friedrich T, Derpsch R. Successful Experiences and Learnings from Conservation Agriculture Worldwide. Online 8th World Congress on Conservation Agriculture (8WCCA). The Future of Farming: Profitable and Sustainable Farming with Conservation Agriculture. Bern (Switzerland); 2021

[93] Lal R. Feeding 11 billion on 0.5 billion hectare of area under cereal crops. Food and Energy Security. 2016b;**5**(4):239-251. DOI: 10.1002/fes3.99

[94] Lal R, Mohtar RH, Assi AT, Ray R, Baybil H, Jahn M. Soil as a basic Nexus tool: Soils at the Center of the Food–Energy–Water Nexus. Current Sustainable/Renewable Energy Reports. 2017;**4**(3):117-129

[95] Lal R. Feeding the world and returning half of the agricultural land back to nature. Journal of Soil and Water Conservation. 2021;**76**(4):75. DOI: 10.2489/jswc.2021.0607A

[96] Naab JB, Mahama GY, Yahaya I, Prasad PVV. Conservation agriculture improves soil quality, crop yield, and incomes of smallholder farmers in North-Western Ghana. Frontiers in Plant Science. 2017. DOI: doi.org/10.3389/fpls.2017.00996

[97] Newton P, Civita N, Frankel-Goldwater L, Bartel K, Johns C. What is regenerative agriculture? A review of scholar and practitioner definitions

based on processes and outcomes. Frontiers in Sustainable Food Systems. 2020;**4**. DOI: 10.3389/fsufs.2020.577723

[98] Ogle S, Alsaker C, Baldock J, Bernoux M, Breidt F, McConkey BG, et al. Climate and soil characteristics determine where No-till management can store carbon in soils and mitigate greenhouse gas emissions. Scientific Reports. 2019;**9**:11665. DOI: 10.1038/s41598-019-47861-7

[99] Pooniya V, Zhiipao RR, Biswakarma N, Jat SL, Kumar D, Parihar CM, et al. Long-term conservation agriculture and best nutrient management improves productivity and profitability coupled with soil properties of a maize–chickpea rotation. Scientific Reports. 2021. DOI: doi.org/10.1038/s41598-021-89737-9

[100] Pittelkow CM, Liang X, Linquist BA, Van Groenigen KJ, Lee J, Lundy ME, et al. Productivity limits and potentials of the principles of conservation agriculture. Nature. 2015b;**517**:365-368

[101] Rose S, Khatri-Chhetri A, Stier M, Wiese-Rozanova L, Shelton S, Wollenberg E. Ambition for Soil Organic Carbon Sequestration in New and Updated Nationally Determined Contributions. Analysis of Agricultural Sub-sectors in National Climate Change Strategies. Wageningen, The Netherlands: CGIAR; 2021

[102] Vermeulen S, Bossio D, Lehmann J, Luu P, Paustian K, Webb C, et al. A global agenda for collective action on soil carbon. Nature Sustainability. 2019;**2**(1):2-4. DOI: 10.1038/s41893-018-0212-z

[103] WCCA. Declaration. The 8th World Congress on Conservation Agriculture 21-23 June. Bern, Switzerland; 2021

[104] Zerihun A, Tadesse B, Shiferaw T, Kifle D. Conservation agriculture: Maize-legume intensification for yield, profitability and soil fertility improvement in maize belt areas of Western Ethiopia. International Journal of Plant and Soil Science. 2014;**3**(8):969-985

Chapter 2

Application of Crop Modeling in Multi-Cropping Systems for Maximize Production and Build Resilient Ecosystem Services

Addisu Ebbisa

Abstract

One of the main challenges in the transition to more sustainable agriculture is designing and selecting agricultural systems that are stable and perturbation resistant. Crop diversification is now recognized as a decisive part of sustainable agroecological development. It is one of the crucial agroecological practices that prove ecosystem services such as nutrient cycling, biological N fixation, pest and disease regulation, erosion control, climate regulation, soil fertility maintenance, biodiversity conservation, and carbon sequestration. To maximize these desired outcomes, understanding, designing, and optimizing, the adoption of crop diversification is crucial for the sustainability of food production under low-input practices. One approach to building sustainable food security and optimal management systems for limited resources is through the application of crop simulation models in multi-cropping systems. Indeed, some models can be used to simulate intercropping systems such as DSSAT, APSIM, ALMANAC, STICS, and FASSET. Thus, the application of such powerful models provides an option to redesign crop mixtures in appropriate sowing proportion and sowing date to tackle the enormous challenges facing agricultural development. In this regard, this review intended to assess existing suitable model to simulate multiple cropping systems and its role in building resilient crop production and ecosystem services without damaging the environment. It also highlights the key role of crop diversity as an ecosystem service provider to guarantee plant productivity in emerging systems of sustainable agriculture.

Keywords: building resilient, multi-cropping, biodiversity, crop simulation models, ecosystem services, crop production, sustainable agriculture

1. Introduction

One of the current emerging challenges in agricultural sectors is to ensure increasing food demand for the growing world population and built resilient (agro)

ecosystem services. Intensive agriculture is often proposed as a solution to feed the growing global population [1] with greater inputs of agrochemicals, water, and others [2], regardless of the environmental consequences [3–5]. However, this in turn can cause another environmental issue (s) such as a reduction in soil quality, loss of biodiversity, disease, and pest resistance, reduction in water quality, and high dependency on fossil fuel-based energy, and/or contribute to greenhouse gas emissions [2, 3, 6–9]. These consequences led to a search for new pathways leading to promoting new strategies that can sustain agriculture production without jeopardizing human health and ecosystem services [2, 10].

One possible approach would be a diversification cropping system that is a key factor in developing a more sustainable agroecological system [1]. A key agroecological principle in agricultural production systems is diversification in time (rotation) and space (intercropping, agroforestry systems, biofertilizers, and cover crops) *via* mechanisms of competition, facilitation, complementarity, and compensation [11], which lies in how much diversified the agro-production system is and the positive interaction among the diverse components of the system [12, 13]. Multispecies cropping systems can significantly reduce fertilizer overuse problems thereby minimizing the environmental impacts of agriculture while maintaining high food production [14]. It proved ecosystem services such as nutrient cycling, biological N fixation, pest and disease regulation, erosion control, climate regulation, maintenance of soil fertility, biodiversity conservation, and carbon sequestration [15, 16]. The agroecological practice of intercropping, meaning the simultaneous cultivation of two or more species in the same field [4] and specific sequences [12, 17], has recently gained renewed interest as a means of ecological intensification [11].

According to Maezieux et al.'s [6] review of agroecosystems, biodiversity may (i) contribute to constant biomass production and reduce the risk of crop failure in unpredictable environments, (ii) restore disturbed ecosystem services, such as water and nutrient cycling, and (iii) reduce risks of pests, and diseases through enhanced biological control or direct control of the pest. To maximize these desired outcomes understanding, designing, and optimizing the adoption of crop diversification the system is crucial for the sustainability of food production under low input practices. Thus, the application of crop modeling is a powerful tool that provides an option to redesign crop mixtures, sowing proportion, plant arrangement, and sowing date and to tackle the enormous challenges facing agricultural development [18]. It is a useful tool in capturing the interactions among climatic conditions, soil types, and nutrient dynamics in cereal-based farming systems and generates knowledge for aiding agricultural developments that would otherwise be impossible through field experimentation [18, 19].

The study by Tsubo et al. [20] demonstrates the possibility of applying a crop simulation model to assess the growth and yield of cereal-legume intercropping over time and space. Carberry et al. [21] and Berghuijs [22] demonstrate the capacity of the Agricultural Production Systems Simulator (APSIM) crop model to simulate competition between species (intercropping) in an agricultural system. APSIM is also able to simulate the soil carbon, water, and nitrogen balances arising from interactions between different crops and pastures grown in rotation [23]. The experiment of Yi-tao et al. [24] suggests that the suitability DNDC model could be used to simulate yield production and N uptake in intercropping systems in the North China Plain. Similarly, Brisson et al. [25] adopt the STICS model to the intercrop model by extrapolation of a sole crop model and concluded that it is useful to evaluate various combinations of crops, including arable crops, forage, and perennial crops. Moreover, Baumann et al. [26] analyzed the competition, crop yield, and plant quality in an intercropping system using an eco-physiological model (INTERCOM). Besides, Berghuijs et al. [27] developed a novel, parameter-sparse process-based crop growth model (Minimalist Mixture Model, M3) to simulate strip intercrops, and proposed that total intercrop yield can be improved by selecting specific traits related to the phenology of both species. Pembleton et al. [28] approve the resilience of forage crops to climate change scenarios as an important component of dairy forage production in southeastern Australia using APSIM crop modeling. In this regard, this study aims to review the existing suitable models applied to simulate multiple cropping systems and their role in building resilient crop production and ecosystem service to feed dramatic population growth without damaging the environment. It also highlights the key role of crop diversity to build resilient crop production and ecosystem services to safeguard food, water, and environmental quality.

1.1 Core principles of agroecology

Agroecosystems are the most intensively managed ecosystems, capable of producing a high harvestable yield through the application of optimal agrochemical and energy management techniques [29] while maintaining the robust vitality of the soil and other environmental conditions [5]. The most commonly used definition of agroecology is an application of ecological concepts and principles for the study, design, and management of sustainable agroecosystems. The application of ecological principles in agriculture is a key part of the global response to climate instability for meeting significant increases in our food needs [30]. The core principles of agroecology are (1) planning and securing the health of the whole system by enhancing beneficial biological interactions and synergisms; (2) minimizing the use of external resources and optimizing the use of nutrients and energy on the farms; and (3) promote agro-biodiversity [31]. These agroecological principles inspire a variety of farming practices such as conservation tillage, crop rotation and fallowing, cover crops, and mulching, mixing crops in a single plot, mixed crop-livestock systems, integrated nutrient management, efficient water harvesting, agroforestry, and holistic landscape management [31]. These practices can help improve soil health and carbon sequestration, water quality and nutrient flows, and control pests and diseases, and they can make farming systems more climate resilient [15]. By reducing dependence on external inputs, agroecology can reduce producers' vulnerability to economic risk and enhance the ecological and socioeconomic resilience community [30]. Setting up multiple cropping systems to maintain crop production while significantly reducing inputs (mineral fertilizers, water, energy, and pesticides) and providing regulation and cultural services requires much more than an understanding of species coexistence and the identification of species functions [4].

1.2 Concept of resiliency and sustainability in the agriculture system

The concepts of production, efficiency, stability, and resilience lie at the heart of natural ecosystem characterization by ecologists. Sustainability in agriculture is the practice that meets the needs of the current generation without compromising the needs of future generations [32, 33] *via* stable, equitable, and profitable applications of ecosystem management practices. The goal of sustainable agriculture is to maximize the net benefits that society receives from the agricultural production of food, fiber, and ecosystem services [33] by maintaining existing productivity and enhancing sustainability. This will require increased crop yields and resource use efficiency based on ecological management practices [33].

The term resilience in agroecology is defined as the greater capacity of an ecosystem to withstand and recover from various forms of stress, including herbivorous pests, diseases, droughts, and floods [3, 30, 34]. The resilience of the crop production system also refers to the largest departure from the optimal conditions that the crop production system can sustain without losing its production capacity [35]. It is constructed or emerges through the aggregation of two or more mutually reinforcing livelihood outcomes [36]. Essentially, resilience is measured in three ways: (1) the amount of change the system can undergo and still retain the same controls on function and structure; (2) the degree to which the system is capable of self-organization; and (3) the ability to build and increase the capacity for learning and adaptation [37]. This confirmed that resilience is related to the ability to ensure and guarantee system functions in the face of economic, social, environmental, and institutional disturbances through robustness, adaptability, and transformability [38]. Increasing the resilience of agricultural livelihoods is key to making sustainable development by monitoring and predicting crisis and disaster risks in the agriculture sector [30].

1.3 Role of plant diversification for resilient ecosystem services

Recent agriculture has minimized diversity in favor of vulnerable monocultures and such systems show intrinsically less stability and resilience to perturbations. Diversity among and within species provides insurance or a buffer against environmental fluctuations because different species and varieties occupy different niches and respond differently to change [36]. Agroforestry, intercropping, conservation agriculture, doubled-up legume cropping, fertilizer micro-dosing, planting basins, and push-pull technology were identified as key agronomic innovations widely promoted in sub-Saharan Africa [39]. These outcomes, in turn, could lead to an increase in the resilience of rural households and communities concerning environmental, socioeconomic, and climatic stresses [36]. The impacts of agroforestry on crop yield, soil quality, and pest control are context-specific and depend on the ecological conditions, the type of tree species, and the type of crop. Because of their deep roots and year-round vegetation cover, agricultural systems with trees and shrubs are inherently more sustainable and efficient in using plant nutrients than annual systems without trees [39]. Promoting the cultivation of leguminous crops, grasses, shrubs, and trees offers multiple advantages, for example, augmenting crop and soil productivity that is adapting to climate change by increasing the resilience of agroecosystems [40].

Diversifying farming systems can provide significant ecological and economic benefits and such as food and nutritional security, income generation, and better health [35, 41]. It is perceived as a strategy to simultaneously achieve high productivity and maintain environmental sustainability [42]. It can also provide a variety of ecosystem services depending on the type (positive, neutral, or negative) and degree of interaction between biodiversity and local environmental conditions, which affect ecosystem functioning as well as the economic status of the community [4]. This kind of diversity can also provide ecosystem services, for example, regulation and control of pests and diseases, rehabilitation of fields with poor soil fertility, reduced soil erosion, sustenance of pollinator diversity, and support of below-ground biodiversity [43]. These benefits could reduce the financial, environmental, and personal health risks that usually result from a high level of (externally sourced) agricultural inputs, which is crucial for achieving global food security [36]. Generally, the main ecosystem services provided by multi-cropping are benefits for crop production (e.g., yield quality, quantity, and stability), improvement of soil biogeochemistry, improvement of biological pest control/management, and climate regulation by mitigating greenhouse gas emissions [13].

1.4 Intercropping and yield stability

Intercropping is a way to increase diversity in an agricultural ecosystem. According to the review of Bedoussac et al. [16], intercropping leads to (i) higher and more stable grain yield (0.33 versus 0.27 kg m^{-2}), (ii) higher cereal protein concentration (11.1 versus 9.8%), (iii) higher and more stable gross margin (702 versus 577€ha^{-1}), and (iv) improved use of abiotic resources according to species complementarities for light interception and use of both soil mineral nitrogen and atmospheric N_2 than mean sole crops. Similarly, intercropping provides insurance against crop failure or unstable market prices for a given commodity [44]; increases food security in vulnerable production systems [45]; and is a feasible entry point to ecological intensification [46]. Thus, it offers greater financial stability than sole cropping, which makes the system particularly suitable for labor-intensive small farms. Besides, intercropping allows lower inputs through reduced fertilizer and pesticide requirements [22], thus minimizing the environmental impacts of agriculture [44, 47]. Thevathasan and Gordon [48] revealed that the tree/crop agroforestry system was four times more C sequestration potential in the fast-growing tree than that of conventional agricultural fields. Because of reduced fertilizer use and more efficient N-cycling, the tree-intercropping systems could also lead to the reduction of nitrous oxide emissions from the agricultural field [4, 5, 15, 16]. This system also increases soil organic carbon content, bird, insect, and earthworm diversity abundance and distribution, which indicates a sustainable land-management option for long term-productivity [43, 48].

Diversity at all levels, from genetics to the ecosystem, enhances the ability to crop systems to overcome and adapt to forthcoming changes [49]. It reduces interspecific competition by enhancing complementarity or facilitation processes thereby improving the exploitation of resources, which in turn reflected in the increase in plant production corresponding to greater efficiency of the agroecosystem as a whole [50]. Along with food safety, biodiversity supports healthy and nutrient-rich diets, enhances the efficiency of agroecosystems, and boosts resilience to changing environmental conditions, climate risks, and socioeconomic challenges [51–53]. Sunflower intercropping with alfalfa proved the most appropriate and stable yields than sole cropping [49]. Similarly, intercropping may contribute to the mitigation of climate change, for example, by reducing the need for fossil-based N fertilizer, mechanical weed control, and the associated N_2O and CO_2 emissions [11]. Legumes used in intercropping and doubled-up legume technology reduce reliance on nitrogen fertilizer and pesticide inputs then lower the GHG emission [39]. Generally, compared to intensive agriculture, intercropping optimizes ecosystem services such as yield stability, utilizes resources efficiently, suppresses pests and diseases, mitigates climate change, controls soil pollution, and increases on-farm biodiversity intercropping through reducing the use of agro-chemicals [5, 54, 55].

2. Crop simulation models (CSM) in multiple cropping systems

2.1 Modeling interspecific competition

Crop simulation models (CSMs) as decision support tools for intercrop/multi-crop systems and future directions for modeling multi-crop systems [56] were developed to simulate soil–plant-atmosphere interactions by considering environmental variables, genotype-specific traits, and their response to the environment using daily through mathematical equations [58] to make research, the management, or teaching more effective [59]. They are useful tools to examine the feasibility of agricultural management systems and can be used to examine the effect of trees within cropping systems [60].

Models dealing with interspecific competition get more and more important as the postulation for sustainable agricultural production has become a global political issue [60]. To model intercropping in terms of neighboring effects in the context of field boundary cultivation, Knörzer et al. [61] developed and integrated a new model approach into the DSSAT model. Different models are considered for modeling interspecific competition in different ways, for example, DSSAT [60, 62], DNDC [24, 55]; ALMANAC [64, 65], APSIM [55, 58], ERIN, FASSET, GAPS, GROWIT, INTERCOM, KMS, NTRM-MSC, SIRASCA, SODCOM, SOYWEED, STICS [66], VCROPS, and WATER-COMP [61]. One main strength of these models is that they consider the effects of several abiotic stresses (e.g., water, N, and temperature) and their interactions on crop performance providing a quantitative estimate at a relevant scale (e.g., yield ha – 1). Similarly, models such as Model Soil, Water, Atmosphere and Plant (SWAP2 × 1D), World Food Studies (WOFOST) [67], CROPSYST (Cropping Systems Simulation Model [57, 68], Daisy, Environmental Policy Integrated Climate (EPIC) and Agricultural Policy/Environmental eXtender (APEX), Agricultural Land Management Alternatives with Numerical Assessment Criteria (ALMANAC) [64, 65], WaNulCAS, and Hi-sAFe are capable of simulating growth and the yield of the crop with response to their environment and management of multiple cropping systems [19, 56, 61]. Among the numerous crop growth models, the most widely used models are the DSSAT and APSIM models. They are potentially relevant for addressing the performance of crop mixtures compared to that of sole crops under a variety of environmental conditions, such as drought or nutrient limitations [61].

Since early studies, two dominant crop simulation model types are mechanistic and empirical [56]. Mechanistic (eco-physiological) or process-based crop models (PBCMs) simulate the growth, development, and performance of crop plants by modeling their underlying physiological processes, and the coordination and integration of these processes at the whole-plant and canopy scales based on "focal plant–neighbor plant" interactions [69, 70]. The physiological processes incorporated into PBCMs can include photosynthesis, transpiration, respiration, organ development, and assimilate transport. For process-oriented models, the turbid layer medium analogy (where the canopy structure is described by statistical distributions.) has proven to be the most useful [61]. The majority of CSMs use the mechanistic approach to model crop systems. On the contrary, empirical (descriptive) models are direct descriptions of observed data used to estimate final yield and are generally expressed as regression equations with one or a few factors. They are useful for making predictions within the range of data used to parameterize them but are not suitable for extrapolation. Such formal description at the logical level may perfectly reflect the properties of a real system in the "entry-exit" terms within a relatively narrow class and a limited range of affecting factors but is almost not associated with the essence of physical, chemical, and biological effects in the soil–plant-atmosphere system example Agricultural Land Management Alternatives with Numerical Assessment Criteria (ALMANAC) model that can only supported by experimental data [69].

2.2 How modeling build resilience multi-cropping system?

2.2.1 Model can provide an opportunity to assess the suitability and sustainability of cropping systems under projected climate change

Cereal-legume intercropping has a substantial impact on enhancing higher yield, yield stability, and food security community [71]. The yield stability of intercropping systems is important in developing cropping systems that produce economic yields in response to variations in the environment due to years and locations [72]. Changing climate adversely affects agricultural productivity and creates food insecurity. Crop growth models are modern and efficient tools that have been extensively used in mounting the climate change impacts and developing adaptation packages for sustainable crop production in changing climate [73]. They suggested that the negative effect of future climate change on maize production systems can be minimized or overcome by modifying the sowing dates and fertilizer (fertigation) and developing heat and drought-tolerant hybrids. A study by Msongaleli et al. [63] approves the applicability of DSSAT and APSIM crop simulation models as tools for assessing possible impacts of climate change on sorghum under projected climate scenarios. Similarly, Chimonyo et al. [74] applied the APSIM model and recommend that changing plant populations and sequential in maize landrace with Bambara groundnut intercropped system increase yield and WUE under projected climate change. This allows for the identification of short-, medium-, and long-term strategies to aid in mitigating the impacts of climate change on productivity and WUE. They stated that crop diversity could enhance crop productivity, stability, and thus food security, through efficient water utilization. Shili [75] test the impact of a living cover crop on the agronomic and environmental performance of the system for different climatic and technical scenarios using the STICS crop model adapted. They found that, in most climatic scenarios, the emergence of the fescue crop during the late tillering phase of the wheat crop gave the best compromise between wheat yield overall nitrogen accumulation and radiation interception. Furthermore, APSIM could adequately simulate expert knowledge, that is, expected yields, and of important crops with adequately simulated competitive effects in maize-bean intercropping systems [76].

2.2.2 Explore water management, GHG regulation, and adaptation strategies

The increase in the concentration of greenhouse gases (GHGs) in the atmosphere has led to an elevated concern and urgency to adopt measures for carbon (C) sequestration to mitigate climate change. Carbon sequestration plays a major role in mitigating climate change by converting atmospheric carbon into long-lived wood biomass and soil carbon pool [77]. Agriculture is the second-largest contributor to GHG emissions in terms of CO_2 eq. contributing about 31% of Africa's total emissions [39]. Most practices can lower emission intensities (e.g., intercropping, conservation, agroforestry) and biomass (e.g., agroforestry) by reducing direct and indirect soil N_2O emissions and by increasing the amount of carbon stored in the soil [39]. Integrating strip intercropping, conservation tillage, as well as straw mulching, significantly boosts crop yields, improves resources use efficiency, and alleviates food security in arid areas while lowering the carbon emissions from farming [78].

Mathematical models have the promising potential to explore solutions to water management problems. Archontoulis et al. [79] test and calibrate APSIM for the crop, soil water, soil N, surface organic matter, manure, and soil temperature and prove to be a reliable model that can be used as a research and decision tool for the agricultural system. Fung et al. [55] stated that DNDC model able to assess the environmental values of intercropping in terms of alleviating air pollution and safeguarding a sustainable food supply. They show that maize-soybean intercropping systems have the potential to save 42% of fertilizer application, compared with their monoculture counterparts, while producing high yields, improving both fertilizer-use and land-use efficiencies. Crop environment resource synthesis (CERES) in DSSAT and world food studies (WOFOST) in SWAP were used to simulate the growth, development, and soil water balance based on field experiment data [80]. Similarly, Hernández-Ochoa et al. [19] identified and tested the robustness of some models (e.g., APSIM, DSSAT, EPIC, STICS, and WOFOST) for the regulation of GHGs by simulating N_2O emissions and mitigation of climate change. Banerjee et al. [62] assess the impact of projected climate on the growth and yield of rice-lentil-groundnut cropping sequence using DSSAT and suggested that rice benefited from preceding groundnut and residue, hence, could sustain the yield in a long term.

Application of agroecosystem modeling can also dynamically simulate a diverse set of regulating and provisioning ecological services such as regulation of greenhouse gas emissions, water quality, and soil erosion [19, 80]. Simulation of water quality by simulating soil N retention (*via* N leaching dynamics) is possible for most models' examples are CropSyst, DNDC, STICS, EPIC, and APSIM [19]. SWAP2 × 1D and WOFOST can simulate the water balance components and crop growth [67]. Likewise, the STICS growth model simulates crop growth and development, as well as water and N balance to improve understanding of interspecific interactions and explore best options of strategies management [75, 81]. This model also paves a possible way to recycle mineral nitrogen efficiently in multiple cropping systems without any effect on water balance and environmental conditions [75]. Moreover, Araya et al. [82] evaluate the impacts of cropping systems and water management on the yield performance of selected dominant cropping systems in the highlands of Africa using DSSAT modeling and highlight the significance of integrating diverse cropping systems (that include legumes) and water management practices (tied ridges and irrigation) for agroecological intensification. Thus, it helps to control the balance between competition and facilitation then improving the agronomic practice for resilient ecosystem service provision in a holistic manner.

2.2.3 For carefully designing and selecting the best adaptive practice

Cereal-legume combinations are known to facilitate the efficient utilization of nutrients by creating a congenial environment [83]. Plant models able to infer plant–plant interactions can be helpful for the identification of major interaction traits and the definition of ideotypes adapted to a targeted intercropping system [84]. The crop simulation model dealt with competition for light and can be used to assess risk for intercrop productivity over time and space in semi-arid regions [20]. Agricultural system models are important tools for understanding complex system interactions to achieve multiple productivities and environmental goals [79]. Models are used extensively for understanding the behavior of the crops in specific environments, and optimization of planting dates, fertilizer application, and crop choice. Multi-cropping systems have potential advantages in productivity, stability of outputs, resilience to disruption, and ecological sustainability [6]. Multispecies systems can also provide other services, linked to the quality of the environment: Trees and cover crops can provide shade and shelter for animals and humans. Although frequent, the advantages and benefits of multispecies systems must not be over-generalized: Not all crops are beneficial in mixtures, since they do not systematically generate ecological and/or economic benefits, and may involve more complex or higher inputs of labor [6]. So, using crop modeling, it is possible to develop an innovative planting design, management practice, and crop varieties for mixed-species plantation [22] through ecological, agricultural, and genetic concepts and approaches [4, 56]. These varieties can modify about criteria of agronomy needs and holistic environmental issues, which lead to higher yields and quality than the corresponding pure crop. Baumann et al. [26] determine ranges of plant densities that enable the intercropping system to meet the current quality standards of the component crops.

Using APSIM modeling, Nelson et al. [85] support the suitability of intercropping to achieve high-yield production or reduce risk under drought and an opportunity to diversify food production. Similarly, APSIM is also used to develop best management practices for improved yield and WUE of sorghum-cowpea intercropping system [86]. DSSAT and APSIM models have been already employed as promising tools to discover likely options for better nitrogen management and water-saving techniques, thereby bringing nitrogen- and water-efficient best management practices to different cropping systems in semi-arid tropics [87]. The results study by Gautam et al. [88] concluded that diversification of rice fallows with the inclusion of short-duration pulses/oilseeds is one of the options to achieve higher profitability, system productivity, and sustainability in the long run. Hoffman [42] proves the usefulness of APSIM model applications for the design of suitable cropping systems in addressing various dimensions of sustainability. They suggest intercropping is a promising option for cropping system diversification.

2.2.4 For optimization of traditional farming systems and yield gap analysis

Traditional farming systems like intercropping or mixed cropping are known to be the embryonic form of sustainable production concerning biodiversity, resource use efficiency, and yield stability [61]. As field trials are time consuming and expensive, models are the alternatives. Agroecosystem models can be used to simulate the basic effects of crop rotation on crop yields, resource use dynamics, and efficiency. Most crop modeling can simulate the performance of intercropping systems in response to the climate and soil conditions and allows the evaluation of management intervention through tillage, irrigation, or fertilization as well as choice, timing, and sequencing of crops such as APSIM [89], STICS [90], and ALMANAC for weed relay intercropping with wheat FASSET, DNDC [24], and INTERCOM [26, 61]. Crop yield simulation is an important component of yield-gap analysis and numerous studies have been published that use simulation models to assess crop yield gaps (quantified as the difference between potential and actual farm yields), the impact of climate change on future crop yields, and land-use change [91]. Modeling can allow for the verification of estimated yield gaps with on-farm data and experiments [92]. Similarly, Rizzo et al. [93] suggest double-cropped soybean cropping systems as an alternative for increasing grain production in the main agricultural region of the world after analyzing their yield gaps.

2.2.5 For improving land use and management

A landscape generator typically considers different agricultural land use systems including natural, semi-natural habitats, cropland, and landscape elements. Maize-cowpea intercropping with a temporal niche difference is a better option for sustainable crop production and maximizing land use [94]. Meixiu et al. [95] showed that intercropping could be used to obtain more yield on less land with less water by developing and application of dynamic process-based modeling taking into account the acquisition of light and water by the component species. Holzkamper et al. [96] determine the ideal configuration of grassland, farmland (without a specific crop specified), and woodlands for particular bird species in Northwest Saxony, Germany, using a spatial optimization model for land use modification tradeoffs between species habitat appropriateness and management.

APEX is being used in the USDA-NRCS CEAP Cropland National Assessment to evaluate the effectiveness of conservation practices, including the impacts of conservation practices on pesticide losses from farm fields. The optimum setup for species habitats and management was provided by smaller patches and greater diversity of land use including more forest lands and de-creased grassland and cropland. Accordingly, EPIC and APEX models are the most flexible and dynamic tools that can be used to estimate the impacts of land management, conservation practices, and/or climate on a wide range of environmental indicators, including water quantity; wind, water, or channel erosion; soil carbon sequestration; pesticide fate and movement; nutrient (nitrogen and phosphorus) cycling and losses *via* surface runoff (both soluble and sediment-bound phases); leaching; volatilization; and tile drainage [65]. Plotkin et al. [97] demonstrate the value and utility of APEX in agricultural fate modeling for evaluating the environmental benefits of conservation practices such as residue management and conservation tillage, as well as identifying areas where conservation practices may be required. This shows APEX model can replicate measured stream flow and sediment yields for rangeland watersheds with satisfactory performance based on well-accepted statistical criteria [98].

3. Summaries and conclusions

Multi-cropping systems and agroecological approaches can improve resource use efficiency for both nutrients and water, thereby facilitating low-input agricultural practice. It can help to develop more sustainable and resilient farming systems that combine stable yields with enhanced biodiversity and ecosystem services to feed a growing world population. To further increase sustainability, there is a need to expand the research to consider other management strategies such as the use of other traditional crop species, fertilization, rainwater harvesting, and soil conservation techniques. A key point in future modeling challenges remains the need for creating bridges between ecophysiology, population biology, and functional ecology. Indeed, some models can be used to simulate intercropping systems. These models often include competition for light, water, and N, such as DSSAT, APSIM, ALMANAC, STICS, and FASSET. Similarly, the DNDC model is also able to simulate yield and N uptake for intercropping systems under different N application rates. Thus, the model can explore soil and water management strategies, GHG regulation, and its adaptation mechanism then can provide an opportunity to assess the suitability and sustainability of cropping systems under projected climate change. Based on these modeling outs, one can design and build more sustainable crop production and resilient ecosystem service for the future generation holistically. Finally, for optimizing adoption and use of intercropping for all stockholders, further scientific development in simulation and awareness creation is urgently required. This should relate to the development of strong ethics for sustainable management of soil, water, and natural resources.

Acknowledgements

The authors are highly thankful to researchers whose findings are included directly or indirectly in preparing this manuscript.

Funding

The authors received no direct funding for this research.

Conflicts of interest

The authors declare no conflict of interest.

Data availability

All data generated are included in this article reference's part.

Author details

Addisu Ebbisa
Faculty of Agriculture, Department of Plant Science (Agronomy), Haramaya University, Dire Dewa, Ethiopia

*Address all correspondence to: saebbisa@gmial.com; aebbisa@gmail.com

References

[1] Dwivedi A et al. Towards sustainable intensification of maize (Zea mays L.) + legume intercropping systems; experiences; challenges and opportunities in India; a critical review. Journal of Pure and Applied Microbiology. 2016;**10**(1):725-740

[2] Alori ET, Babalola OO. Microbial inoculants for improving crop quality and human health in Africa. Frontiers in Microbiology. 2018;**9**:1-12. DOI: 10.3389/fmicb.2018.02213

[3] Malézieux E. Designing cropping systems from nature. Agronomy for Sustainable Development. 2012;**32**:15-29. DOI: 10.1007/s13593-011-0027-z

[4] Gaba S, Lescourret F, Boudsocq S, Enjalbert J. Multiple cropping systems as drivers for providing multiple ecosystem services: From concepts to design. Agronomy for Sustainable Development. 2015;**35**:607-623. DOI: 10.1007/s13593-014-0272-z

[5] Yang H, Zhang W, Li L. Intercropping: Feed more people and build more sustainable agroecosystems. Frontiers of Agricultural Science and Engineering. 2021;**8**(3):373-386

[6] Maezieux E et al. Mixing plant species in cropping systems: Concepts, tools, and models. A review. Agronomy for Sustainable Development. 2009;**29**:43-62

[7] Bhardwaj D, Ansari MW, Sahoo RK, Tuteja N. Biofertilizers function as a key player in sustainable agriculture by improving soil fertility, plant tolerance, and crop productivity. Microbial Cell Factories. 2014;**13**(66):1-10

[8] Teklewold H, Kassie M, Shiferaw B, Köhlin G. Cropping system diversification, conservation tillage and modern seed adoption in Ethiopia: Impacts on household income, agrochemical use and demand for labor. Ecological Economics. 2013;**93**:85-93. DOI: 10.1016/j.ecolecon.2013.05.002

[9] Wimalawansa SA, Wimalawansa SJ. Agrochemical-related environmental pollution: Effects on human health. G.J.B.A.H.S. 2014;**3**(3):72-83

[10] Javaid A. Role of Arbuscular Mycorrhizal fungi in nitrogen fixation in legumes. In: Khan MS, Musarrat J, Zaidi A, editors. Microbes for Legume Improvement. Germany: Springer Wien New York; 2010. pp. 409-426

[11] Jensen ES, Chongtham IR, Dhamala NR, Carton N, Carlsson G. Diversifying European agricultural systems by intercropping grain legumes and cereals. International Journal of Agriculture and Natural Resources. 2020;**47**(3):174-186. DOI: 10.7764/ijanr.v47i3.2241

[12] Sau S, Sarkar S, Das A, Saha S, Datta P. Space and time utilization in horticulture based cropping system: An income doubling approach from the same piece of land. Journal of Pharmacognosy and Phytochemistry. 2017;**6**(6):619-624

[13] Gaudio N et al. Current knowledge and future research opportunities for modeling annual crop mixtures. A review. Agronomy for Sustainable Development. 2019;**39**(20):1-20

[14] Mudare S, Kanomanyanga J, Jiao X, Mabasa S, Lamichhane JR. Yield and fertilizer benefits of maize/grain legume intercropping in China and Africa: A meta-analysis. Agronomy for Sustainable

Development. 2022;**42**(81):1-17. DOI: 10.1007/s13593-022-00816-1

[15] Wezel A et al. Agroecological practices for sustainable agriculture. Agronomy for Sustainable Development. 2014;**34**(1):1-20. DOI: 10.1007/s13593-013-0180-7

[16] Bedoussac L et al. Ecological principles underlying the increase of productivity achieved by cereal-grain legume intercrops in organic farming. A review. Agronomy for Sustainable Development. 2015;**35**:911-935. DOI: 10.1007/s13593-014-0277-7

[17] Martin-Guay M, Paquette A, Dupras J, Rivest D. The new green revolution: Sustainable intensification of agriculture by intercropping. Science of the Total Environment. 2018;**615**:767-772. DOI: 10.1016/j.scitotenv.2017.10.024

[18] Whitbread AM, Robertson MJ, Carberry PS, Dimes JP. How farming systems simulation can aid the development of more sustainable smallholder farming systems in southern Africa. European Journal of Agronomy. 2010;**32**:51-58. DOI: 10.1016/j.eja.2009.05.004

[19] Hernández-Ochoa IM et al. Model-based design of crop diversification through new field arrangements in spatially heterogeneous landscapes. A review. Agronomy for Sustainable Development. 2022;**42**(74):1-25

[20] Tsubo M, Walker S, Ogindo HO. A simulation model of cereal – Legume intercropping systems for semi-arid regions I. model development. Field Crops Research. 2005;**93**:10-22. DOI: 10.1016/j.fcr.2004.09.002

[21] Carberry PS et al. Application of the APSIM cropping systems model to intercropping systems. Australian Journal of Experimental Agriculture. 1996;**36**:1037-1048. DOI: 10.1071/EA9961037

[22] Berghuijs HNC et al. Calibrating and testing APSIM for wheat-faba bean pure cultures and intercrops across Europe. Field Crops Research. 2021;**264**(108088):1-14. DOI: 10.1016/j.fcr.2021.108088

[23] Carberry PS et al. Simulation of a legume ley farming system in northern Australia using the agricultural production systems simulator. Australian Journal of Experimental Agriculture. 1996;**36**:1037-1048

[24] Yi-tao Z, Jian LIU, Hong-yuan W, Qiu-Liang LEI, Hong-bin LIU, Li-mei Z. Suitability of the DNDC model to simulate yield production and nitrogen uptake for maize and soybean intercropping in the North China plain. Journal of Integrative Agriculture. 2018;**17**(12):2790-2801. DOI: 10.1016/S2095-3119(18)61945-8

[25] Brisson N et al. Adaptation of the crop model STICS to intercropping. Theoretical basis and parameterization. Agronomy for Sustainable Development. 2004;**24**(6-7):409-421. DOI: 10.1051/agro

[26] Baumann DT, Bastiaans L, Goudriaan J. Analysing crop yield and plant quality in an intercropping system using an eco-physiological model for interplant competition. Agricultural Systems. 2002;**73**:173-203

[27] Berghuijs HNC, Wang Z, Stomph TJ. Identification of species traits enhancing yield in wheat-faba bean intercropping: Development and sensitivity analysis of a minimalist mixture model. Plant and Soil. 2020;**455**:203-226

[28] Pembleton KG, Cullen BR, Rawnsley RP, Harrison MT. Modelling

the Resilience of Forage Crop Production to Future Climate Change in the Dairy Regions of Southeastern Australia Using APSIM. Cambridge University Press; 2016. pp. 1131-1152. DOI: 10.1017/S0021859615001185

[29] Thomas V, Kevan P. Basic principles of Agroecology and sustainable agriculture. Journal of Agricultural and Environmental Ethics. 1993;**1**:1-19. DOI: 10.1007/BF01965612

[30] FAO. Guiding the Transition to Sustainable Food and Agricultural Systems: The 10 Elements of Agroecology. FAO; 2018

[31] Kazakova Y, Radeva D. The Role of Agroecosystems Diversity towards Sustainability of Agricultural Systems. Sofia, Bulgaria: EAAE; 2015

[32] Hooker JE, Black KE. Arbuscular Mycorrhizal fungi as components of sustainable soil-plant systems. Critical Reviews in Biotechnology. 1995;**15**(3/4):201-212

[33] Tilman D, Cassman KG, Matson PA, Naylor R, Polasky S. Agricultural sustainability and intensive production practices. Nature. 2002;**418**:671-677

[34] Gabella JI, Strijker D. Sustainability or resilience ? A case study in the semi-arid Pampean region of Argentina. Resilience. 2019;7(1):1-20. DOI: 10.1080/21693293.2018.1446298

[35] Zampieri M, Weissteiner CJ, Grizzetti B, Torti A, Van Den Berg M, Dentener F. Estimating resilience of crop production systems: From theory to practice. Science of the Total Environment. 2020;**735**:139378. DOI: 10.1016/j.scitotenv.2020.139378

[36] Vernooy R. Does crop diversification lead to climate-related resilience? Improving the theory through insights on practice improving the theory through insights on practice. Agroecology and Sustainable Food Systems. 2022;**46**(6):877-901. DOI: 10.1080/21683565.2022.2076184

[37] Cabell JF, Oelofse M. An indicator framework for assessing agroecosystem resilience. Ecology and Society. 2012;**17**(1):1-13

[38] Cleves A, Youkhana E, Toro J. A method to assess agroecosystem resilience to climate variability. Sustainability. 2022;**14**(8588):1-26

[39] Kuyah S et al. Innovative agronomic practices for sustainable intensification in sub-Saharan Africa. A review. Agronomy for Sustainable Development. 2021;**41**(16):1-21

[40] Meena RS, Das A, Singh G, Lal R. Legumes for Soil Health and Sustainable Management. Vol. 1. Singapore: Singapore Pte Ltd; 2018

[41] Glaze-Corcoran S, Hashemi M, Sadeghpour A, Herbert SJ. Understanding intercropping to improve agricultural resiliency and environmental sustainability. In: Advances in Agronomy. 1st ed. Vol. 162. Elsevier Inc.; 2020. pp. 199-256

[42] Hoffman MP et al. Simulating medium-term effects of cropping system diversification on soil fertility and crop productivity in southern Africa. European Journal of Agronomy. 2020;**119**:1-15. DOI: 10.1016/j.eja.2020.126089

[43] Brillouin D, Ben T, Malézieux E, Seufert V. Positive but variable effects of crop diversification on biodiversity

and ecosystem services. Global Change Biology. 2021;**00**:1-14. DOI: 10.1111/gcb.15747

[44] Lithourgidis AS, Dordas CA, Damalas CA, Vlachostergios DN. Annual intercrops: An alternative pathway for sustainable agriculture. Australian Journal of Crop Science. 2011;**5**(4):396-410

[45] Mousavi SR, Eskandari H. A general overview on intercropping and its advantages in sustainable agriculture. Journal of Applied Environmental and Biological Sciences. 2011;**1**(11):482-486

[46] Rusinamhodzi L, Corbeels M, Nyamangara J, Giller KE. Maize–grain legume intercropping is an attractive option for ecological intensification that reduces the climatic risk for smallholder farmers in central Mozambique. F. Crop Research. 2012;136:12-22. DOI: 10.1016/j.fcr.2012.07.014

[47] Tahat MM, Alananbeh KM, Othman YA. Soil health and sustainable agriculture. Sustainability. 2020;**12**(4859):1-26. DOI: 10.3390/su12124859

[48] Thevathasan NV, Gordon AM. Ecology of tree intercropping systems in the northern temperate region: Experiences from southern Ontario, Canada. Agroforestry Systems. 2004;**61**:257-268

[49] Babec B, Šeremešic S, Rajkovic M, Hladni N, ´Cuk N, Stanisavljevic D. Potential of sunflower-legume intercropping: A way forward in sustainable production of sunflower in temperate climatic conditions. Agronomy. 2021;11(238):1-18. DOI: 10.3390/ agronomy11122381

[50] Duchene O, Vian J, Celette F. Intercropping with legume for agroecological cropping systems: Complementarity and facilitation processes and the importance of soil microorganisms. A review. Agriculture, Ecosystems and Environment. 2017;**240**:148-161. DOI: 10.1016/j.agee.2017.02.019

[51] Wiebe K, Robinson S, Cattaneo A. Climate Change, Agriculture and Food Security: Impacts and the Potential for Adaptation and Mitigation. Rome, Italy: Elsevier Inc.; 2019

[52] Amoak D, Luginaah I, Mcbean G. Climate change, food security, and health: Harnessing Agroecology to build climate-resilient communities. Sustainability. 2022;**14**(13954):1-15

[53] Muralikrishnan L, Padaria RN, Choudhary AK, Dass A, Elansary HO. Climate change-induced drought impacts, adaptation and mitigation measures in semi-arid pastoral and agricultural watersheds. Sustainability. 2022;**14**(6):1-18

[54] Mazzafera P, Favarin JL, de Andrade SAL. Editorial: Intercropping Systems in Sustainable Agriculture. Frontiers in Sustainable Food Systems. 2021;**5**(634361):1-4. DOI: 10.3389/fsufs.2021.634361

[55] Fung KM, Tai APK, Yong T, Liu X, Lam H. Co-benefits of intercropping as a sustainable farming method for safeguarding both food security and air quality. Environmental Research Letters. 2019;**14**:044011

[56] Chimonyoa VGP, Modi AT, Mabhaudhi T. Perspective on crop modeling in the management of intercropping systems. Archives of Agronomy and Soil Science. 2015;**1**:37-41. DOI: 10.1080/03650340.2015.1017816

[57] Alagappan S. Integrated approaches in crop simulation modeling for future agriculture. Biotech Research Today. 2020;**2**(11):1170-1173

[58] Keating BA et al. An overview of APSIM, a model designed for farming systems simulation. European Journal of Agronomy. 2003;**18**:267-288

[59] Mbabaliye T, Wojtkowski PA. Problems and perspectives on the use of a crop simulation model in an African research. Experimental Agriculture. 1994;**30**:441-446

[60] Huda AS, Ong CK. Crop simulation models and some implications for agroforestry systems. In: The Application of Meteorology to Agroforestry Systems Planning and Management. Vol. 2017. 1989. pp. 1-13

[61] Knörzer H, Graeff-Hönninger S, Müller BU, Piepho H-P, Claupein W. A modeling approach to simulate effects of intercropping and interspecific competition in arable crops. The International Journal of Information Systems and Social Change. 2010;**1**(4):44-65. DOI: 10.4018/jissc.2010100104

[62] Banerjee S, Chandran SM, Mukherjee A, Nanda MK, Kumari VV. Evaluating the long - term impact of projected climate on rice - lentil - groundnut cropping system in lower Gangetic plain of India using crop simulation modeling. International Journal of Biometeorology. 2022;66:55-69. DOI: 10.1007/s00484-021-02189-8

[63] Msongaleli B, Rwehumbiza F, Tumbo SD, Kihupi N. Sorghum yield response to changing climatic conditions in semi-arid Central Tanzania: Evaluating crop simulation model applicability. Agricultural Sciences. 2014;**5**:822-833. DOI: 10.4236/as.2014.510087

[64] Perkins SA, Mankin KD, Nelson R. Modeling the Economic Feasibility of Sweet Sorghum in Western Kansas and the Panhandles of Texas and Oklahoma. Louisville, Kentucky: ASABE; 2011;7004(11)

[65] Alliance AC, Gassman P, Baffaut C. EPIC and APEX: Model use, calibration, and validation. The American Society of Agricultural and Biological Engineers. 2013;**55**(4):1447-1462. DOI: 10.13031/2013.42253

[66] Vezy R et al. Modelling intercrops functioning to advance the design of innovative agroecological systems. Research Square. 2022;**preprint**:0-31

[67] Pinto VM, Van Dam JC, Van Lier QDJ. Intercropping simulation using the SWAP model: Development of a 2 × 1D algorithm. Agriculture. 2019;**9**(126):1-19

[68] Plentinger MC, de Vries FWTP. Rotation Models for Ecological Farming. DLO. Wageningen: Research Institute for Agrobiology and Soil Fertility; 1997

[69] Badenko VL, Topaj AG, Yakushev VV, Mirschel W, Nendel C. Crop models as research and interpretative tools. Plant Biology. SAVCHENKO, Ed. Moscow, Russia: NPO;2017;52(3, I. V):437-445

[70] Kim S-H, Hsiao J. Modelling sub-systems. In: Boote K, editor. Advances in Crop Modeling for Sustainable Agriculture. 2nd ed. Vol. xii. USA: Burleigh Dodds Science Publishing; 2020. pp. 3-43

[71] Raseduzzaman M. Intercropping for Enhanced Yield Stability and Food Security. Faculty of Landscape

Architecture, Horticulture and Crop Production Science, SLU; 2016

[72] Dapaah HK, Asafu-Agyei JN, Ennin SA, Yamoah C. Yield stability of cassava, maize, soya bean, and cowpea intercrops. The Journal of Agricultural Science. 2003;**140**:73-82. DOI: 10.1017/S0021859602002770

[73] Yasin M, Ahmad A, Khaliq T, Habib M, Niaz S. Climate change impact uncertainty assessment and adaptations for sustainable maize production using multi-crop and climate models. Environmental Science and Pollution Research. 2022;**29**:18967-18988. DOI: 10.1007/s11356-021-17050-z

[74] Chimonyo VGP, Wimalasiri EM, Kunz R, Modi AT, Mabhaudhi T. Optimizing traditional cropping systems under climate change: A case of maize landraces and Bambara groundnut. Frontiers in Sustainable Food Systems. 2020;**4**(562568):1-19. DOI: 10.3389/fsufs.2020.562568

[75] Shili I et al. Does intercrop winter wheat (Triticum aestivum) with red fescue (Festuca rubra) as a cover crop improve agronomic and environmental performance ? A modeling approach to cite this version: HAL id: Hal-01173222 crop improve agronomically and environment. Field Crops Research. 2019;**116**(3):218-229. DOI: 10.1016/j.fcr.2009.11.007

[76] Dimes J, Abebe TM, Tefera AT, Nhantumbo N. Evaluation of APSIM to simulate maize-bean cropping systems in eastern and southern Africa: An alternative approach. SIMLESA. 2011;**1**:1-5

[77] Kumar KSN, Maheswarappa HP. Carbon sequestration potential of coconut based cropping systems under integrated nutrient management practices. Journal of Plantation Crops. 2019;**47**(2):107-114. DOI: 10.25081/jpc.2019.v47.i2.5776

[78] Hu F, Gan Y, Cui H, Zhao C, Feng F, Yin W. Intercropping maize and wheat with conservation agriculture principles improve water harvesting and reduces carbon. European Journal of Agronomy. 2016;**74**:9-17. DOI: 10.1016/j.eja.2015.11.019

[79] Archontoulis SV, Miguez FE, Moore KJ. Evaluating APSIM maize, soil water, soil nitrogen, manure, and soil temperature modules in the Midwestern United States. Biometry, Modeling and Statistics. 2014;106(3):1025-1140. Doi:10.2134/ agronj2013.0421

[80] Ines AVM, Droogers P, Makin IW, Das Gupta A. Crop Growth and Soil Water Balance Water Modeling to Explore Water Management Water Options. Colombo, Sri Lanka; 2001

[81] Launay M, Brisson N, Satger S, Hauggaard-Nielsen H, Corre-hellou G, Kasynova E. Exploring options for managing strategies for pea– Barley intercropping using a modeling approach. European Journal of Agronomy. 2009;**31**:85-98. DOI: 10.1016/j.eja.2009.04.002

[82] Araya A, Prasad PVV, Ciampitti IA, Jha PK. Using crop simulation model to evaluate the influence of water management practices and multiple cropping systems on crop yields: A case study for Ethiopian highlands. Field Crops Research. 2021;**260**(November 2020):108004. DOI: 10.1016/j.fcr.2020.108004

[83] Maitra S, Ray DP. Enrichment of biodiversity, influence in microbial population dynamics of soil and nutrient utilization in cereal-legume intercropping systems: A review.

International Journal of Biological Sciences. 2019;**6**:11-19. DOI: 10.30954/2347-9655.01.2019.3

[84] Louarn G, Barillot R, Combes D, Escobar-gutiérrez A. Towards intercrop ideotypes: Non-random trait assembly can promote overyielding and stability of species proportion in simulated legume-based mixtures. Annals of Botany. 2020;**126**:671-685. DOI: 10.1093/aob/mcaa014

[85] Nelson WCD, Hoffmann MP, Vadez V, Rötter RP, Koch M, Whitbread AM. Can intercropping be an adaptation to drought ? A model- based analysis for pearl millet – cowpea. Journal of Agronomy and Crop Science. 2021;**00**: 1-18. DOI: 10.1111/jac.12552

[86] Grace V, Chimonyo P, Modi AT. Applying APSIM for Evaluating Intercropping under Rainfed Conditions: A Preliminary Assessment. 2019

[87] Dubey PK. Increasing resilience in crops for future changing environment. In: Adaptive Agricultural Practices. Vol. 2014. Switzerland; 2020. pp. 45-61

[88] Gautam P et al. Alteration in agronomic practices to utilize rice fallows for higher system productivity and sustainability. Journal of Agronomy and Crop Science. 2021;**260**(108005):1-11. DOI: 10.1016/j.fcr.2020.108005

[89] Wang E et al. Development of a generic crop model template in the cropping system model APSIM. European Journal of Agronomy. 2002;**18**:121-140

[90] Justes E, Roche R, Mary B, Ripoche D. An overview of the crop model STICS an overview of the crop model. European Journal of Agronomy. January 2003;**0301**:309-332. DOI: 10.1016/S1161-0301(02)00110-7

[91] Grassini P et al. How good are good enough ? Data requirements for reliable crop yield simulations and yield-gap analysis. Journal of Agronomy and Crop Science. 2015;**177**:49-63. DOI: 10.1016/j.fcr.2015.03.004

[92] Van Ittersum MK, Cassman KG, Grassini P, Wolf J, Tittonell P, Hochman Z. Yield gap analysis with local to global relevance — A review. Journal of Agronomy and Crop Science. 2013;**143**:4-17. DOI: 10.1016/j.fcr.2012.09.009

[93] Rizzo G, Pablo J, Ernst O. Cropping system-imposed yield gap: Proof of concept on soybean cropping systems in Uruguay. Journal of Agronomy and Crop Science. 2021;**260**(November 2020):107944. DOI: 10.1016/j.fcr.2020.107944

[94] Suhi MA et al. How does maize-cowpea intercropping maximize land use. Land. 2022;**11**(581):1-18. DOI: 10.3390/land11040581

[95] Meixiu T et al. Dynamic process-based modeling of crop growth and competitive water extraction in relay strip intercropping: Model development and application to wheat-maize intercropping. Journal of Agronomy and Crop Science. 2020;**246**:107613. DOI: 10.1016/j.fcr.2019.107613

[96] Holzkämper A, Lausch A, Seppelt R. Optimizing landscape configuration to enhance habitat suitability for species with contrasting habitat requirements. Ecological Modelling. 2006;**198**(3-4):277-292. DOI: 10.1016/j.ecolmodel.2006.05.001

[97] Plotkin S, Wang X, Potter TL, Bosch DD, Williams JR, Hesketh ES, et al. APEX calibration and validation of water and herbicide transport under U.S. southern Atlantic coastal plain

conditions. Transactions of the ASABE. 2013;**56**(1):43-60. DOI: 10.13031/2013.42589

[98] Wang X, Amonett C, Williams JR, Wilcox BP, Fox WE, Tu MC. Rangeland watershed study using the agricultural policy/environmental eXtender. Journal of Soil and Water Conservation. 2014;**69**(3):197-212. DOI: 10.2489/jswc.69.3.197

Chapter 3

Evaluation of the Role of Small-Scale Farmers in Soil and Water Conservation Management in the Context of Climate Change

Tirivashe Phillip Masere

Abstract

The global land resource is increasingly under pressure due to both anthropogenic and natural factors such as unsustainable land management practices and climate change, respectively. Land degradation and climate change are among the major global threats to the resilience of agro-ecosystems and stability of food production systems. Small-scale resource-constrained farmers, who account for the majority of farmers across the world, are the hardest hit due to the scale of their operations, operating environment, and circumstances. Despite these global challenges, small-scale farmers have continued to adjust their farming systems to withstand the vagaries of climate change, while at the same time aiming to achieve land degradation neutrality. This chapter sought to evaluate the role played by small-scale farmers in soil and water conservation management in attempt to address land degradation and climate change. Further, the chapter investigated key characteristics and circumstances of small-scale farmers as well as their constraints, strengths, and opportunities. The chapter argues that farmers' indigenous knowledge system has been and continues to be a key strength and offers an opportunity for which more specialized scientific and agricultural extension support can build upon in developing lasting solutions to climate change and soil and water conservation management.

Keywords: soil and water conservation, small-scale farmers, semi-arid regions, climate change, climate smart agriculture, indigenous knowledge, land degradation neutrality

1. Introduction

Land degradation and climate change are among the major global threats to the resilience of agro-ecosystems and stability of food production systems. Globally, 3.2 billion people are affected by land degradation, most of whom are the small-scale farmers residing in rural communities of Africa [1]. These African small-scale farmers generally practice rain-fed crop and livestock production, which makes them vulnerable to climate change and droughts [2, 3]. The effects and impacts of climate change are felt the hardest in small-scale farming systems of Africa due to the scale of operation, biophysical conditions of their farms, and operating circumstances of the farmers particularly their poor resource endowment [4]. However, life has to go on for these farmers. They have to produce enough food to feed their families and communities, despite the aforementioned challenges. To do this, they have to build resilience of their farming systems to withstand the effects of climate change and to better manage their natural resources (soil and water) as they aim to achieve land degradation neutrality (LDN).

Worldwide there are approximately 570 million farms, 470 million of which are small-scale farms [5]. Of these, approximately 33 million are located in Africa, thus constituting 80% of all farms on the continent [2, 6]. Due to them being the overwhelming majority for both Africa and the world at large, small-scale farmers are uniquely placed at the center of it all—bearer of the brunt of the climate and land degradation challenges on the one hand and on the other, an integral part of efforts to address or manage these challenges. However, this does not seem to be the case, particularly for the African small-scale farmers who are often overlooked and misunderstood in terms of their importance in finding lasting solutions to these land resource challenges threatening their very livelihoods. Yet, for effective implementation of agricultural policies, there is need for the inclusion and incorporating features of small-scale farmers, as agricultural policy and statutory instruments affect different farming sectors differently [7].

It is for this reason that this chapter sought to evaluate the role played by small-scale farmers in soil and water conservation management in attempts to address the impacts and effects of land degradation and climate change to their farming systems. To achieve this goal, it is imperative to start by understanding the small-scale farming sector, its key characteristics, strengths, and constraints. To this end the chapter attempts to achieve five objectives, namely: defining small-scale farmers; determining the characteristics and circumstances of small-scale farming systems in Africa; exploring constraints and strengths of small-scale farmers; evaluating the soil and water conservation strategies employed by small-scale farmers in adapting to climate change; and assessing the land degradation neutrality in small-scale farming systems.

The chapter utilized both primary and secondary data. Secondary data in the form of a review of literature, primarily journals and published technical reports pertaining to small-scale farming systems, were conducted. Conversely, primary data were gathered through focus group discussions (FGDs) with African small-scale farmers, observations, and key informant interviews (KIIs) conducted with public extension agents from Zimbabwe, Zambia, and South Africa.

This chapter proceeds as follows. Section 2 explores the definition of small-scale farmers, including the various factors that have been used in defining the small-scale farmers. Further, key attributes or characteristics of small-scale farmers are also examined. Section 3 evaluates the numerous constraints confronting small-scale farmers due to their scale of operation. Despite all the challenges small-scale farmers face, they still have their strengths, which have kept them going for generations—these strengths are also outlined and discussed in this section. The penultimate section, Section 4 deals with leveraging on the strengths and opportunities of small-scale farmers in adapting to climate change and achieving land degradation neutrality. The final section offers some concluding remarks.

2. Definition, characteristics, and circumstances of small-scale farmers

2.1 Misconceptions about small-scale farmers

A lot of misconceptions arose about the African small-scale farmers due to their circumstances and operating conditions compared to their counterparts in developed world. Among the notable misconceptions, small-scale farmers are often viewed as backward and unproductive farmers operating in native lands [8]. This is further from the truth as small-scale farmers are among to be the main contributors to national food security in most developing countries such as Zimbabwe and Zambia [9]. Further, in most African countries such as South Africa and Zimbabwe, small-scale farmers are generally associated with the black population, while large-scale farmers are generally associated with the white population [8]. These misconceptions seem to stem from the history of land tenure systems under Apartheid regime, and before Independence in South Africa and Zimbabwe, respectively, where blacks were overcrowded in native/rural areas characterized with infertile soils as opposed to the fewer whites who owned prime extensive agricultural lands. Thus, the political and historical development of the Zimbabwe particularly during the 90 years of colonial and settler government also shaped farming systems [10].

Prior to the fast-track land reform program (FTLRP) of the year 2000 in Zimbabwe, there were three distinct farming subsectors: communal lands, resettlement areas, and large-scale commercial farming [10]. The first two subsectors constituted the small-scale farming systems. These were located in areas known as Reserves, which were characterized by poor agricultural potential and were merely established as a labor pool for the white commercial farming community [11].

The FTLRP introduced two farm models, namely: A1 (small self-contained farms) and A2 (slightly larger-scale or medium-scale), focusing on subsistence and more commercial production, respectively [12]. Thus, after the implementation of the FTLRP, a tri-modal structure of Zimbabwe's farming systems ensued namely small-scale farms, and medium-scale commercial and large-scale estates [12]. The Zimbabwean small-scale farming sector currently comprises old resettlement areas (settled before 2000), communal lands, and the A1 farms.

The foregoing indicates that it is inadequate to use only one or two factors to define or categorize farming sectors. There is need to properly define and constitute small-scale farmers in a manner that encompasses all the other key factors. The next two parts of this section attempted to do achieve that.

2.2 Defining small-scale farms/farmers

While it is almost impossible to have a universally accepted definition of small-scale farmers, most attempts to understand these farmers and their farms have used certain attributes or criteria including land size, resources, and income. Of these farm size is probably the most obvious and easily used criterion to define small-scale farmers [5, 13]. However, the World Bank views small-scale farmers as those with cropland sizes less than 2 hectares and with the added condition of having little or no assets [14]. A similar definition for small-scale farmers is given in a study by FAO, which emphasizes limited/stretched resources in comparison with medium- and large-scale farmers [15]. The limited resources include land holding sizes, livestock, access to inputs and markets, agro-ecological factors, level of technology use, and income levels [10, 13].

Although farm size is the mostly used indicator to identify or define small-scale farmers, it is not a good criterion and has its limitations [8, 16]. As such small-scale farms should not be generalized as simply scaled-down versions of large-scale farms [8]. One reason for this is that land qualities vary across different types of farms, implying that a small land area of high fertility will likely result in a higher yield output compared to a much larger land area characterized by low fertility [16]. As such land size is not a proper determinant of farm-scale categories. Computing net farm profitability would be a much better determinant [8, 16]. Europe has been using this approach in categorizing farms.

Standard gross margins (SGMs) of all farm enterprises are undertaken to evaluate the farm's potential income generation and total farm profitability [7]. In this way, the SGM tool provides information about the scale of a farm's business. The SGM tool can also be adjusted to cater for farms in different localities. This approach is a move-away from using farm area and intensity of production in classifying farms. This is one way to go about defining small-scale farms and farmers. However, it may present two main challenges particularly for small-scale farmers in Africa. Firstly, profitability is not the primary purpose for farming, household consumption and food security are. Secondly, it ignores the non-monetary aspects and attributes that cannot be easily expressed in monetary terms. As such, it is submitted that there is need for inclusion of other characteristics and circumstances in defining small-scale farmers, most of which are discussed under the next subheading.

2.3 Characteristics and circumstances (constraints and strengths) of small-scale farmers

Several factors frame the characteristics of farmers of any scale. Under this subheading, key unique characteristics and circumstances of small-scale farmers are discussed. From these, some constraints, strengths, and opportunities within the small-scale farming sector can be discerned:

2.3.1 Primary purpose

The main objective of small-scale farming systems is for household consumption as opposed to large-scale commercial farming systems that respond to market demand [17]. In all the FGDs conducted with small-scale farmers in Zimbabwe and Zambia, farmers highlighted that while their main reason for farming was for household consumption (**Table 1**). If and when there is a surplus, it will be sold to generate income for their other needs.

2.3.2 Land tenure and size issues

The majority (85%) of small-scale farmers in Africa operate in farms of less than 2 hectares [2]. In addition to their already small farm sizes, African small-scale farmers are also challenged by the ever rising populations, which are absorbed into farming leading to further shrinkage of their farm sizes. This fragmentation of small-scale farms to accommodate high population growth in Africa confines small-scale farmers to subsistence crop production. In addition to small land sizes, small-scale farmers are typically located on lands with marginal production potential (degraded lands). Furthermore, lack of clear and sound system of land rights transfer has been noted as one of the causes of food insecurity and underdeveloped agriculture [3].

Crop type	Crop grown	Reasons for growing
Cereals	Maize	Household consumption, income generation from selling surplus. Stock feeding.
	Sorghum	Household consumption and income generation from selling surplus. Beer brewing for selling and traditional ceremonies.
	Rapoko	Beer brewing for selling and traditional ceremonies.
Legumes	Groundnuts	Household consumption and income generation from selling surplus. Fixing nitrogen into the soil.
	Sugar beans	Household consumption, income generation from selling surplus. Fixing nitrogen into the soil.
	Cowpeas	Household consumption, income generation from selling surplus. Fixing nitrogen into the soil.
	Round nuts	Household consumption, income generation from selling surplus. Fixing nitrogen into the soil.
Tubers	Sweet potatoes	Household consumption and income generation from selling surplus.
	Potatoes	Household consumption.
Vegetables	Including butternuts, onions, tomatoes, cabbage, spinach, chomolia, tsunga, pumpkins, carrots, tomatoes.	Income generation and household consumption.

Table 1.
Crops grown by small-scale farmers and reasons for growing them (source: FGDs).

2.3.3 Semi-arid environment and over reliance on rain-fed agriculture

Most of small-scale farmers in Africa are located in semi-arid environments with low, erratic rainfall (with high prevalent of severe dry spells) and high temperatures—further limiting crop productivity [2, 10, 18]. Further, these semi-arid environments are characterized by marginal and infertile soils [10]. These conditions make meaningful crop production very difficult even for drought resistant crops such as sorghum [10]. Climate change is expected to worsen these poor conditions resulting in crop yield reductions leading to severe food security challenges especially so if household food requirements are not met [2, 3].

2.3.4 Self-reliance

Farmers'self-reliance relates to the extent to which they depend on their capacity and capability as reflected by their knowledge, skills and labor, or lack thereof, to take charge of the factors that affect their farm operations [19, 20]. This entails that small-scale farmers are going to utilize knowledge/information/training available to them as they go about their operations. According to respondents in this study, the majority of small-scale farmers are not formally trained in agriculture. The only training they have is from the public extension workers who often target progressive small-scale farmers and provide them with relevant farming information and technologies, which they are expected to disseminate to other farmers [21]. The training includes planting methods, and soil and water conservation techniques, among other technologies.

Lack of access to agricultural information, particularly farm management information, is a common characteristic of small-scale farmers [22, 23]. In Zimbabwe, they are particularly reliant on extension workers, other farmers through their farmers' clubs and, to a small extent, radio [18, 22, 23]. This was also confirmed during FGDs and KIIs. The public extension workers (via KIIs) in all the three countries acknowledged that farmers self-organize into farmers' clubs whose responsibilities include conducting field days, which offers an informal platform where small-scale farmers exchange experiences and agricultural information through open discussions. This farmer-to-farmer extension can also be an important way to disseminate and encourage adoption of new technology [18].

2.3.5 Family labor

Small-scale farmers and their families provide labor requirements to meet all the farm operations including land preparation, cultivation, weeding, and harvesting [13, 24, 25]. A large family consisting of able-bodied members is thus more likely to be successful compared to a smaller family or families consisting mainly of young children and the aged [17]. In most cases, small-scale farmers and their family are willing to and actually invest more energy and time in their farms than those justified at standard market wage rates because the rewards accrue directly to the family [17]. Both FGDs and KIIs across study sites indicated that small-scale farmers do not place a monetary value to the labor they put into their own farms because they do not perceive it as a cost. This again speaks to the primary purpose for farming, discussed above, that most small-scale farmers are mostly growing crops and rearing livestock for household consumption.

2.3.6 Technology paradigm

As already discussed in the preceding subheading above, small-scale farmers have often relied on their indigenous knowledge, family labor, and influence of social networks on technology adoption for all their farm operations. Use of simple farming technologies is common among small-scale farmers. In both the FGDs and KIIs, it was noted that most small-scale farmers of Southern Africa own hoes, ox-drawn plow, axes, wheelbarrow, and cultivators. The better-off small-scale farmers additionally own scotch carts, harrows, rippers, and ridgers.

Cattle are the main source of power for tilling the land and other farm operations [11, 23]. However, about 40% of the small-scale farmers do not own cattle and must hire them for the required operations [23]. This is consistent with FGDs' findings, where farmers highlighted that some farmers owning implements but without cattle enter into reciprocal cooperative arrangements with farmers who have cattle but lack some implements for tillage and cultivation purposes.

Social function (processes and systems) influences farmer decision making around technology adoption. Small-scale farmers may actually adopt an unfavorable technology to them only because it is preferred by his/her social referent group [26]. In this manner, "social influence" can thus be taken to mean the extent to which members of a referent group affect one another's behavior and experience social pressure to perform particular behaviors.

2.3.7 Use of indigenous knowledge

Most small-scale farmers depend on their own indigenous knowledge generated through many years of farming experience in their own communities to guide crop management decisions [18, 22, 23, 25] (**Table 2**). Although reliable, indigenous knowledge keeps farmers operating at low levels of productivity [11]. This reliance on indigenous knowledge-driven methods of farming had been inaccurately taken to mean that small-scale farmers are mostly uneducated, illiterate, and backward. In this study, it was observed that most small-scale farmers were literate and educated at least to the basic primary school level. Their continued reliance on indigenous knowledge systems (IKS) is mainly because of lack of proven alternatives and lack of access to modern technologies.

Despite the lack of access to modern technologies, small-scale farmers have shown a willingness to learn about modern farming technologies if given the opportunity [18, 23]. This is consistent with FGDs' findings where farmers indicated that whenever their lives are at stake, they are ready to learn and try out new methods or technology when resources are permitting.

IKS technology	Description of the IKS method/technology	Reason and/ advantages of the IKS method
Harvest and seed preservation	Use of fire smoke to preserve dried maize cobs of high yielding local open pollinated varieties (OPVs) to be used a seed next season.	This technology is easy to operate and there are no costs involved, instead they save the cost of buying hybrids.
	Burning gumtree (eucalyptus) leaves and cow dung to repel weevils inside the granaries.	The burning is aimed at eliminating oxygen in the granary to ensure no weevils will survive. This is a no-cost technology to farmers.
	Mixing paraffin and ash for treating and preserving cowpeas seed from weevils,	It is a low cost and easy to implement technology
Crop protection	Use of sand soil and donkey manure to control pest and diseases in field crops, for example control of maize stalk borer.	No-cost technology to farmers.
Seasonal rainfall forecasting	Studying local indigenous indicators like fruiting of certain indigenous tree species, position of the moon, wind direction and behavior of birds to indicate a "good" and "poor" rainfall season.	The indigenous indicators are more reliable in predicting the nature of rainfall season than the official/scientific seasonal climate forecast. The indigenous is readily available to farmers unlike the scientific technology which they may not get on time or at all.
Soil fertility	Mixing poultry droppings with water to form what small-scale call "chicken soup" which they use as a top dress fertilizer	Easy to implement and use. It is also a no-cost technology to farmers.
Multiple cropping and intercropping	Growing multiple crops in one field for example cover crops and runner crops like pumpkins to ensuring total ground cover.	It minimizes the impact of raindrops and thus controls soil erosion.
Tree and hedge planting (Live fencing)	Live fencing for marking homestead and field boundaries and protecting crops from straying animals through planting of trees, shrubs and hedges.	The live fence also acts as windbreaks which reduce wind velocity and hence erosion by wind. Vegetation binds the soil making it less vulnerable to soil erosion.
Spot irrigation	Applying water to the immediate areas around a plant only as opposed to the whole garden.	The technology is water use efficient technology which saves water especially in poor rainfall seasons.

Table 2.
Technology developed by farmers through their indigenous knowledge systems (IKS).

2.3.8 Production paradigm

Small-scale farming systems in sub-Saharan Africa are characterized by low yields as a result of the low level of production they usually operate at. Average crop yields in small-scale farming systems are usually very low and sometimes fail to meet the household requirements or income needs due to persistent droughts, poor soils, lack of good quality inputs, limited or no access to credit, and extension services [23, 25, 27]. Further, small-scale farmers normally use local resources in their farming operations although they may occasional make use of external inputs [2, 13]. Moreover, the small-scale farmers employ risk-averse strategies and aim to maximize yields from constraining resources [28].

2.3.9 Lack of access to credit facilities and markets

Lack of access to credit is the most critical resource constraint to small-scale farmers [11, 22, 23]. The majority of small-scale farmers are unable to access credit from banks and micro-credit firms due to lack of collateral [3]. As a result, these farmers rely on their own meager savings and remittances, thus thwarting any meaningful attempts to expanding their farm productivity [3]. Lower Gweru farmers indicated during FGDs that lack of access to credit facilities one of the major reasons why they have not adopted modern technologies that have high initial costs.

Lack of access to markets is twofold: firstly, lack of access to markets to acquire inputs, and, secondly, lack of access markets to sell their produce. The input and output markets are either missing or incomplete. This presents another challenge—higher transaction costs [8]. Thus, small-scale farmers often fail to use quality inputs due to inaccessibility and high costs [21]. Lack of a ready market to sell produce has resulted in large post-harvest losses in Africa [29]. Post-harvest losses in sub-Saharan Africa are estimated to amount to more than 40% and are even as high as 70% for perishables [30].

Closely related to access to markets are the infrastructural circumstances for most small-scale farmers. A poor road network hinders smooth distribution of inputs to the farms as well as output (agricultural produce) from the farms to the market. Due to poor road systems in small-scale farming areas of Zimbabwe and most developing sub-Saharan Africa farmers resort to inefficient modes of transportation such as animal-drawn scotch carts [3].

2.3.10 Mixed farming systems/integrated crop-livestock systems

Most small-scale farming systems are characterized as mixed farming systems, comprising both crop production and livestock production. Small-scale farmers often grow staple crops such as maize, sorghum, groundnuts and also rearing cattle, goats, and poultry [23]. FGD respondents across the three countries indicated that they mostly grow maize, sunflower, groundnuts, finger millet, and sorghum. These crops are also usually intercropped.

There are numerous advantages of this mixed farming systems including reduction of risk normally prevalent in a monoculture farming system [2]. Although not every small-scale farmer in Zimbabwe, Zambia, and South Africa owns cattle, the majority of them have goats, sheep, and poultry. Most of these farmers give priority to the crop production over livestock production [31]. This is mainly because they want to meet their household food security requirement first. This section discusses the advantages of the integrated crop-livestock systems.

Source of income: The livestock production enterprise presents opportunities for regular income generation including the selling of products such as milk and meat. Combining this income to that generated seasonally or regularly from the crop production enterprise improves the cash flow of the farmers [32]. Poor small-scale farmers with goats can meet short-term immediate needs for cash and meat. Farmers with cattle can also generate income through hiring out their cattle as draft power to those without for draft power purposes [24, 31]—this was observed and confirmed through FGDs and KIIs in the three study countries. Cash from livestock sales is also used for meeting other important household needs including education, family health, and acquiring farming inputs.

Draft power: Despite the majority of small-scale farmers not having tractors, those with cattle can use their cattle for draft power to pull implements for their farming operations [24, 31]. While farmers without cattle may be able to hire draft power from those who own cattle, they are inconvenienced as they will have to wait until the owners finish their own farm operations first, leading to delayed land preparations and planting [24].

Food security enhancement: Milk and meat produced from the livestock production enterprise enhance the household food security while simultaneously improving the general nutrition status of households [31]. Additionally, the income generated from selling meat and milk that can be used to supplement food supplies when necessary.

Spreading agricultural risk: Integrated crop-livestock production systems provide an insurance against the risk of total failure. If part or all of the crop production enterprises fail, farmers will fall back on the livestock enterprise, and vice versa.

Synergies between crop and livestock enterprises: In the FGDs, respondents highlighted that in addition to crop residues, they also reserve some grains for livestock and poultry feeding. In turn, the livestock and poultry provide manure, which farmers used in their fields. As already discussed, these farmers rely mostly on organic fertilizer (cattle and poultry manure and mulches) and rarely acquire inorganic fertilizers due to their low input and risk aversion nature. The income to purchase such fertilizers is again generated from the sale of livestock products and surplus crop produce. FGDs' respondents as well as KII findings highlighted those farmers who practiced a technique known as mobile kraals experienced significant increases in crop yield. The technique is also noted for improving soil fertility through cow dung and urine [33].

3. Leveraging on farmers' strengths and opportunities in adapting to climate change and achieving land degradation neutrality

Small-scale farmers, despite all the challenges they face, still have their strengths, which have kept them going for generations. These include use of indigenous knowledge, family labor, low cost of production, integrated crop and livestock systems, spreading of agricultural risk, and conservation of natural resources. These strengths can be leveraged and built upon in the quest to enhance the resilience of farming systems to adapt and better cope with climate change and land degradation. Equally some constraints, already discussed in the previous section, can also be viewed as opportunities. For example, the lack of cattle for tillage purposes has become an opportunity for adoption of conservation agriculture. Further, some small-scale farmer constraints present opportunities for scientists, extension agencies, technology developers, government, and other key stakeholders to work with farmers in introducing relevant interventions. The strengths and opportunities of small-scale farmers and their systems toward soil and water conservation and achieving land degradation neutrality are discussed in detail below:

3.1 Soil and water conservation in small-scale farming systems

Small-scale farmers are usually better at conserving and managing natural resources through their traditional multiple cropping systems (beans, corn, potatoes, and fodder), which reduce soil erosion [34]. Soil erosion is a major form of land degradation and is very costly in that it involves loss of organic matter and nutrients from the soil and the deposition of such nutrients in receiving waters such as rivers and dams where it presents other costly off-sites problems. Multiple cropping reduces yield losses caused by weeds, pests and diseases, and utilizes water, radiation, and nutrients more efficiently, thus resulting in yield advantages of between 20 and 60%, thereby contributing as much as 20% of global food supply [34].

Other indigenous knowledge systems used by respondent farmers to improve soil fertility and water conservation include mulching, composting, animal manure, intercropping, use of crop residues to cover the soil, and use of anthill and ashes to improve the soil structure and to lime the soil, respectively. Most these indigenous technologies are closely linked to the climate smart agriculture technologies, which have been promoted by most public extension agencies and nongovernmental organizations (NGOs) in Africa. Technologies such as conservation agriculture and thermal composts are highly adopted by small-scale farmers due to their having traits similar to farmers' own indigenous practices such as *gatshombo* (planting basins) and makeshift composts made from crop residues and grass, respectively [23]. Respondents felt they owned these technologies although they acknowledged the technologies had been upgraded and improved by experts. This emphasizes the need for scientists, extension agencies, and technology developers to build on farmers' indigenous knowledge and experiences in coming up with relevant interventions to challenges affecting farmers.

Conservation agriculture was highly adopted mainly because farmers found it to increase crop yields, reduce soil erosion, and use water and nutrients efficiently as application of these resources will be done in the planting area only. In a study across 13 districts in Zimbabwe, crop yield increases of up to 300% were observed for three seasons (2004/05 up to 2006/07) [35]. More important to the most small-scale farmers, conservation agriculture does not require draft power to establish. As such it was very popular among farmers with fewer or no cattle. However, it was noted that even farmers who owned cattle also adopted it. For these farmers (who owned cattle), their other option for soil moisture conservation was deep tillage of fields a couple of months before the onset of the rainy season to increase permeability and thus water absorption capacity during the rainy season.

Similarly, thermal compost technology was highly adopted because for a number of reasons. Firstly, it is less costly to implement and use. Secondly, it was considered a locally available option to mineral fertilizer and cattle manure (particularly for farmers owning few or no cattle). Finally, and perhaps most importantly, it improved crop yields at similar rates to mineral fertilizers. According to FGDs, the use of compost also helps to reduce leaching of nutrients from the soil. Further, soil fertility is maintained or improved by using composts as opposed to using inorganic fertilizers, which some farmers in Zimbabwe argued that it hardens the soil.

3.2 Toward land degradation neutrality in small-scale farming systems

Land degradation is taking place at unprecedented levels, contributing to a dramatic decline in the land productivity throughout the world. Further, when land degradation occurs, soil carbon and nitrous oxide are emitted from the soil to that released into the atmosphere, thus making land degradation one of the most important contributors to climate change [1, 33]. About 24 billion tons of fertile soil is estimated to be lost annually, largely because of unsustainable agriculture practices, and if this trend continues unabated, 95% of the Earth's land areas could become degraded by 2050 [1]. Hence, there is need for concerted efforts to arrest and reverse land degradation in all its forms as acknowledged in Agenda 2030 Sustainable Development Goals Target 15.3, which states the need to strive for a land-degradation-neutral world.

Land degradation neutrality [LDN] is defined as "a state whereby the amount and quality of land resources necessary to support ecosystem functions and services and enhance food security remain stable or increase within specified temporal and spatial scales and ecosystems" [36]. It is concerned about managing land more sustainably to reduce degradation, while increasing rates of land restoration. The two ends (reducing degradation and land restoration) converge to give a zero-net rate of land degradation [33, 36].

As already discussed, small-scale farming is among the worst affected by land degradation due their farms' biophysical conditions and locality. As such they also have a role to play in ensuring attainment of land degradation neutrality at their level. In the previous sections, it was discussed how small-scale farmers, through their own indigenous knowledge system and practices, are managing their natural resources. Among them are the multiple cropping systems, intercropping, and no-tillage practices, which ensure total soil cover to reduce soil erosion and surface runoff. Such actions can contribute toward achieving land degradation neutrality.

Some of the constraints of small-scale farmers have indeed become opportunities, as already discussed. The lack of adequate livestock and money for hiring cattle for purposes of conventional tillage purposes have meant increased adoption of the no-till and reduced and minimum tillage strategies. These strategies align well with the three tenets of conservation agriculture, which are as follows: no or minimum mechanical soil disturbance (through no-till seeding); maintenance of soil mulch cover (with crop residues, stubbles, and cover crops); and diversified cropping (involving annuals and perennials, including legumes, in sequences/rotations).

The benefits of implementing conservation agriculture include limiting of greenhouse gases such as carbon dioxide, as carbon is kept in the soil where it is needed for crop production as opposed to being emitted into the atmosphere. This provides huge ecological and economic benefits in the fight against climate change. Thus, the use of conservation agriculture as a means to achieve land degradation neutrality has the potential to contribute to the attainment of other related SDGs, for instance, poverty eradication (SDG 1), food security (SDG 2), water (SDG 6), and climate change (SDG 13). This is the reason why LDN is considered as an SDG accelerator, which offers cost-effective and ecological sound means of meeting these goals [33].

4. Conclusions

The chapter suggests that small-scale farmers who are in the majority across Africa and the world are the most hit by global environmental challenges such as climate change and land degradation due to their scale of operation, circumstances, and the biophysical conditions of their farms. As such, they are uniquely placed to play an important role in the development of lasting solutions to the land degradation and climate change. However, this has not been the case as small-scale farmers are often ignored and misunderstood, and their farming systems are often deemed backward and unproductive. The chapter thus attempted to define small-scale farmers, their key characteristics, strengths, and opportunities as well as how these may be leveraged on in adapting their systems to climate change effects and impacts and achieving land degradation neutrality. The chapter outlined that small-scale farmers have been managing their natural resources (soil and water) through their indigenous knowledge systems and practices, most of which aligns well with the three interlinked principles of conservation agriculture (no or minimum mechanical soil disturbance; maintenance of soil mulch cover; and diversified cropping). These farmers may not know or fully comprehend the potential scientific and ecological benefits and implications of some of their tried and tested indigenous practices toward reversing, reducing, and avoiding land degradation and climate change. This then offers gaps and opportunities for scientists and researchers to build capacity of the farmers and perfect some of their indigenous technologies. Thus, instead of ignoring, trivializing, and wrongly perceiving them as backward and unproductive, there is need to engage small-scale farmers and embracing their indigenous knowledge systems and practices as they are uniquely placed to do their part in the contributing toward addressing global challenges, which threatens their very livelihoods. The chapter acknowledges that more work still needs to be done in Southern Africa with regard to the actual assessment of land degradation neutrality using the land restoration indicators: land cover; land productivity; and carbon bank/stock. The numerous economic and ecological benefits of implementing conservation agriculture and related variants of climate smart agriculture, toward attainment of land degradation neutrality (SDG 15), poverty eradication (SDG 1), and climate change (SDG 13), were noted.

Acknowledgements

The author is grateful to the respondents for their valuable time, which they gave generously to provide data for this study.

Conflict of interest

The author declares no conflict of interest.

Author details

Tirivashe Phillip Masere
Department of Land and Water Resources Management, Midlands State University, Gweru, Zimbabwe

*Address all correspondence to: maseretp@staff.msu.ac.zw

References

[1] Global Environment Facility. Land Degradation. 2019. Available from: https://www.thegef.org/sites/default/files/publications/gef_land_degradation_bifold_2019.pdf. [Accessed: September 8, 2022]

[2] Altieri MA, Koohafkan P. Enduring Farms: Climate Change, Smallholders and Traditional Farming Communities. Penang: Third World Network; 2008

[3] Salami A, Kamara AB, Brixiova Z. Smallholder agriculture in East Africa: Trends, constraints and opportunities. Tunis: African Development Bank; 2010;**105**:1-51

[4] Moyo S. Using market research to inform product development: The case of small farmer financial products in South Africa. Agrekon. 2002;**41**(2):189-194

[5] Lowder SK, Skoet J, Raney T. The number, size, and distribution of farms, smallholder farms, and family farms worldwide. World Development. 2016;**87**:16-29. DOI: 10.1016/j.worlddev.2015.10.041

[6] Nagayets O. Small farms: Current status and key trends: Information brief. In: Proceedings of the Future of Small Farms Research Workshop. United Kingdom: Wye College; 2005

[7] FAO. Agenda 10: Asia and Pacific commission on agricultural statistics. In: 23rd Session Siem Reap. Cambodia: FAO; 2010

[8] Kirsten JF, van Zyl J. Defining small-scale farmers in the south African context. Agrekon. 1998;**37**(4):551-562

[9] High Level Panel of Experts. Investing in Smallholder Agriculture for Food Security. 2013. Available from: http://www.fao.org/fileadmin/user_upload/hlpe/hlpe_documents/HLPE_Reports/HLPEReport-6_Investing_in_smallholder_agriculture.pdf. [Accessed: September 9, 2022]

[10] FAO. Zimbabwe's natural regions and farming systems: Fertilizer use by crop in Zimbabwe. Land and plant nutrition management service. Land and Water Development Division. 2006. Available from: https://www.fao.org/3/a0395e/a0395e06.htm#bm06. [Accessed: September 9, 2022]

[11] Mushunje A. Farm Efficiency and Land Reform in Zimbabwe (Unpublished Doctoral Thesis). Alice: University of Fort Hare; 2005

[12] ScoonesI,MarongweN,MavedzengeB, MurimbarimbaF,MaheneheneJ,SukumeC. Zimbabwe's Land Reform: A Summary of Findings. Brighton: IDS; 2011

[13] Chamberlin J. Defining smallholder agriculture in Ghana: Who are smallholders, what do they do and how are they linked with markets? In: Prepared as Part of the Ghana Strategy Support Program and Submitted for Consideration as an IFPRI Discussion Paper. Accra: International Food Policy Research Institute; 2007

[14] World Bank. Reaching the Rural Poor: A Renewed Strategy for Rural Development. Washington, D.C.: World Bank; 2003

[15] Dixon J, Taniguchi K, Wattenbach H. Approaches to assessing the impact of globalization on African smallholders: Household and village economy modeling. Rome: FAO; 2003

[16] Lund P, Price R. The measurement of average farm size. Journal of Agricultural Economics. 1998;**49**(1):100-110

[17] Delgado C. Sources of growth in smallholder agriculture in sub-Saharan Africa. Agrekon. 1999;**38**(1):165-189

[18] Masere TP, Worth S. Applicability of APSIM in decision making by small-scale resource-constrained farmers: A case of Lower Gweru communal area, Zimbabwe. Journal of International Agricultural and Extension Education. 2015;**22**(3):20-34

[19] Worth S. The facilitated learning agenda and the extension carousel. In: Paper Presented at the Annual Conference of the Association for International Agricultural and Extension Education. Maimi, FL: AIAEE; 2014. DOI: 10.5191/jiaee.2014.21207

[20] Rottger A. Strengthening Farm-Agribusiness Linkages in Africa: Summary Results of Five Country Studies in Ghana, Nigeria, Kenya, Uganda and South Africa. Rome: Food and Agriculture Organization of the United Nations; 2004

[21] Pazvakavambwa S, Hakutangwi MBK. Zimbabwe's Agricultural Revolution Revisited. Harare: University of Zimbabwe Publications; 2006

[22] Masere TP, Worth S. Influence of public agricultural extension on technology adoption by small-scale farmers in Zimbabwe. South African Journal of Agricultural Extension. 2021;**49**(2):25-42. DOI: 10.17159/2413-3221/2021/v49n2a12785

[23] Masere TP, Worth SH. Factors influencing adoption, innovation of new technology and decision-making by small-scale resource-constrained farmers: The perspective of farmers in Lower Gweru, Zimbabwe. African Journal of Food, Agriculture, Nutrition and Development. 2022;**22**(3):20013-20035. DOI: 10.18697/ajfand.108.20960

[24] Masere TP. The Applicability of the Agricultural Production Systems Simulator (APSIM) Model to Decision-Making in Small-Scale, Resource-Constrained Farming Systems: A Case Study in the Lower Gweru Communal Area, Zimbabwe (Unpublished master's thesis). Pietermaritzburg: University of KwaZulu-Natal; 2011

[25] Masere TP, Duffy KJ. Factors cost effectively improved using computer simulations of maize yields in semi-arid sub-Saharan Africa. South African Journal of Agricultural Extension. 2014;**42**(2):39-50

[26] Ghane F, Samah BA, Ahmad A, Idris K. The role of social influence and innovation characteristics in the adoption of integrated Pest management (IPM) practices by paddy farmers in Iran. International Proceedings of Economics Development and Research. 2011;**5**:217-220

[27] Makhathini S. Opportunities and challenges faced in promoting smallholder farming as an element in rural economic development. In: The Case of Buffalo City Municipality in the Eastern Cape, South Africa (Unpublished master's thesis). Alice: University of Fort Hare; 2013

[28] Scoones I. The economic value of smallholder farmers in the communal areas of southern Zimbabwe. Agricultural Systems. 1992;**39**:339-359

[29] Kamara AB, Mafusire A, Castel V, Kurzweil M, Vencatachellum D, Pla L. Soaring food prices and Africa's vulnerability and responses: An update. Tunis: African Development Bank. 2009;**97**:1-36

[30] UNIDO. Food Processing Pilot Centres: An Approach to Productive Capacity Building for Trade and Poverty Alleviation in Africa. Vienna: UNIDO; 2007

[31] Homann S, van Rooyen A. Unexploited agricultural growth: The case of crop–livestock production Systems in Zimbabwe. African Association of Agricultural Economists (AAAE) Conference Proceedings. 2007:503-506

[32] Ngongoni NT, Mapiye C, Mwale M, Mupeta B. Factors affecting milk production in the smallholder dairy sector of Zimbabwe. Livestock Research for Rural Development. 2006;**18**:72

[33] UNDP. Achieving Land Degradation Neutrality for People and Planet. 2017. Available from: http://www.undp.org/publications/achieving-land-degradation-neutrality. [Accessed: September 6, 2022]

[34] Altieri MA. Small Farms as a Planetary Ecological Asset: Five Key Reasons Why we Should Support the Revitalization of Small Farms in the Global South. Penang: Third World Network; 2008

[35] Twomlow S, Rohrbach D, Rusike J, Mupangwa W, Dimes J, Ncube B. Spreading the word on fertilizer in Zimbabwe. In: Mapiki A, Nhira C, editors. Land and Water Management for Sustainable Agriculture. Proceedings of the EU/SADC Land and Water Management Applied Research and Training Programme. Malawi Institute Management: Lilongwe; 2007

[36] Global Mechanism of UNCCD. Land Degradation Neutrality: The Target Setting Programme. Bonn, Germany: Global Mechanism of UNCCD; 2016

Chapter 4

Range Expansion of *Catha edulis*: Implications on Plant Communities in Upland Zimbabwe

Evelyn Ngarakana, Clemence Zimudzi, Shakkie Kativu and Brita Stedje

Abstract

Invasive plants have had significant impacts on vegetation communities of Zimbabwe. A study was undertaken to determine current and potential distribution of *C. edulis* in Zimbabwe using DIVA GIS and MAXENT, and to determine climatic conditions under which the species thrives, together with. The species population structure and its impact on native species. Results indicate that the species has its highest occurrence frequency in Manicaland Province, followed by Matebeleland South Province. Some 13% occurrence points were recorded at an altitude less than 600 m, 21% at an altitude ranging from 600–999 m, 43% at an altitude between 1000 and 1399 m and 23% at an altitude above 1400 m. *C. edulis* was recorded in areas of maximum temperature range of 34°C and a minimum of 20°C. The species also occurred in regions with a mean precipitation range as low as 60–300 mm and as high as 1000–1261 mm. Further, *C. edulis* distribution is predicted to expand in the Eastern Highlands (Manicaland), parts of Mazowe and Bindura (Mashonaland Central Province) and parts of Matobo (Matebelaland South Province). Diameter class distributions showed an inverse J-distribution in control sites and in all three sampled sections. An irregular bell-shaped distribution was recorded for co-occurring species on *C. edulis* occupied sites. It was concluded that *C. edulis'* regeneration potential is high and that of competing native species is unstable and has the potential to expand beyond the currently occupied sites.

Keywords: *Catha edulis*, predicted distribution, population structure, impacts, Zimbabwe

1. Introduction

Currently, global biodiversity is being threatened greatly by climate change and invasive species [1]. Several dozens of species and variants (including invasives) have been introduced into Zimbabwe [2], a majority of them within the Eastern Highlands where the climate supports the highest plant diversity. These include some suspected variants of *Catha edulis* (Vahl) Forssk. exEndl. *C. edulis* naturally occurs in the horn of Africa down to southern Africa and Madagascar. Its centre of origin is believed to be Ethiopia and Kenya [3].

Observations on spatial distribution of *C. edulis* in Zimbabwe over the past few years indicate that it is spreading, and where this occurs, few other plant species thrive [4]. Immigrants from East Africa are suspected to have introduced different variants of *C. edulis* to Southern Africa [5]. Some such variants have become more aggressive and currently occupying forest margins in Manicaland, Zimbabwe (Eastern Highlands), an area which forms part of the eastern Africa biodiversity hotspot [6]. The Eastern Highlands ecosystem provides freshwater and other ecosystem services to a significant number of people in the region [6].

The climatic conditions under which *C. edulis* in Zimbabwe thrives and its current and potential sites of distribution in the country are not known. That, together with the population structure of the species, requires an urgent investigation as the results may be useful in predicting the species distribution trends and in wildlife management. An assessment of the impact of *C. edulis* on biodiversity of the Eastern Highlands biodiversity hotspot is also needed. Population structure, which partly reflects age and size structure, is indicative of the health and survival capacity of a species [7]. Important life stages of a species can be revealed and aid in wildlife management [8]. Species diversity and evenness make up species composition and the higher the species richness and productivity are the more resilient the ecosystem is [9].

The present study sought to map current and potential distribution of *C. edulis* in Zimbabwe using MAXENT and DIVA GIS, establish climatic conditions under which the species thrives, determine the population structure of *C. edulis* and assess its impact on indigenous species within Vumba Forest area of Zimbabwe. Maximum entropy (MAXENT), which relies only on presence data and background environmental information, was the preferred assessment method. The method also accommodates small sample size and allows for gaps in records [10].

2. Materials and methods

2.1 Study area

This study was conducted in Zimbabwe. About 4% of the country receives an annual rainfall of >1000 mm, and low mean annual temperatures which range from 15 to 18°C. Approximately 32% of the country has annual rainfall below 500 mm, with a high mean annual temperature of 21–25°C [11]. Some 16% of the country is under protection, and includes endemic and critically endangered plant and animal species [12].

Population structure studies were carried out in Vumba, Eastern Highlands of the country (**Figure 1**). Vumba is about 246 km^2 with its highest altitude being 1911 m [14]. The mean yearly precipitation is 1800 mm and majority occurs between November and August [6]. Soils are deep and well weathered [6]. Its vegetation comprises miombo woodland which favours high rainfall, evergreen Afromontane forests and montane grassland [14]. Most of the vegetation types have been exposed to severe disturbances which has paved way for the encroachment of such invasive species as *Cestrum aurantiacum* Lindl., *Lantana camara* L., *Vernonanthura polyanthus* (Spreng.) Vega & Dematteis and *Solanum mauritiunam*

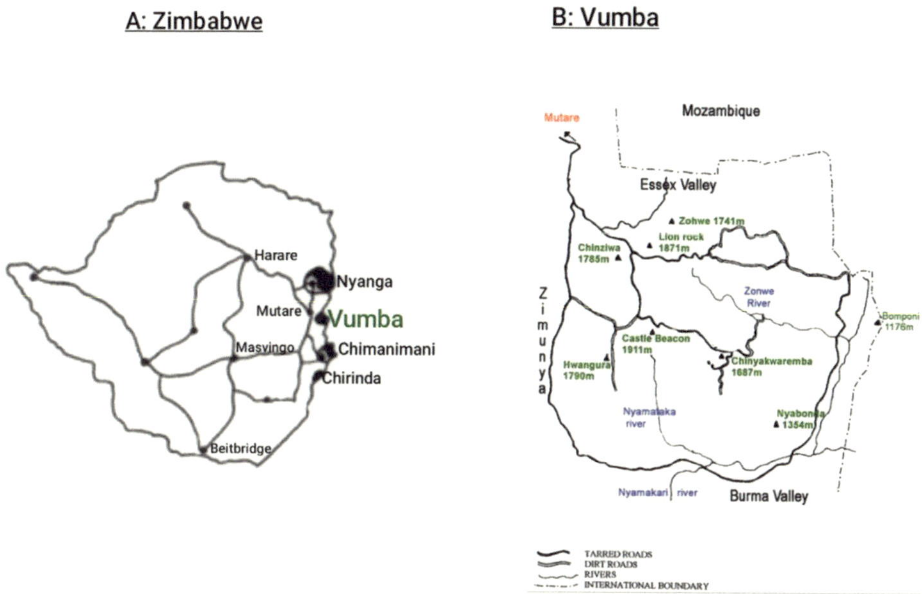

Figure 1.
Study area map in Vumba (right), and its location within Zimbabwe (left) Ballings & Wursten [13].

Scop. [6]. Wattle and eucalypts are also planted for commercial purposes [6]. The following GPS coordinates are of the first, second and third study plots respectively: 19.064640°S, 32.720713°E; 19.078418°S, 32.750628°E; 19.071590°S, 32.744625°E.

2.2 Mapping of *C. edulis* populations

Sources for the species occurrence data were the Zimbabwean National Herbarium and Botanic Gardens, Global Biodiversity Information Facility (GBIF) (www.gbif.org) and Zimbabwe Flora (www.zimbabweflora.co.zw). In the course of field studies, new distribution records of *C. edulis* were recorded. The combined data were checked for duplication. The source for climate variables was WorldClim (www.worldclim.org). WorldClim has a total of 19 bioclimatic variables, 11 temperature and 8 precipitation matrices which represent different annual trends, seasonality and extreme environmental conditions [15]. Pearson correlation coefficients were examined in order to check all the variables for multicollinearity [16]. The first principal component analysis (PCA) analysed only one variable from each set of highly correlated variables ($r > 0.95$). The main bioclimatic variables were also determined by performing the Correlation analysis (CA) and PCA using the statistical program R version 3.1.3 [16]. Spatial resolution t used was 30arc seconds (about 1 km^2 per pixel). This allowed for maximum details [17]. DIVA-GIS [18] was used to map the distribution of *C. edulis* in the whole of Zimbabwe. It is a geographic information system that has been used in various mapping studies such as the mapping of spatial distribution of *Jatropha curcas* L. in Malaysia by Shabanimofrad *et al.* [19] and that of *Senecio vulgaris* L. in China by Cheng and Xu [15]. The software was used according to the manual guide by Scheldeman *et al.* [20].

2.3 Predicting *Catha edulis* potential sites of distribution

MAXENT model was used to predict the potential sites of distribution of *C. edulis*. One-time Split method was used to partition the occurrence records for use in validating the accuracy of the model's predictions [21]. 75% was randomly selected to make up the training data (calibration data) and the remaining 25% was the test data (evaluation data) [17]. Jackknife test was used to measure the importance of the climatic factors. MAXENT was run in default settings. Area Under Curve (AUC) was used to evaluate the predictive ability of the model generated by MAXENT [15]. Results were imported and visualised in DIVA-DIS. The software was used according to the manual guide by Scheldeman *et al.* [20] and explanations by Phillips *et al.* [22].

2.4 Population structure

The study was carried out in October and November 2020. Three *C. edulis* occupied sites representing at least 20% of the total area with the species were randomly selected. Three adjacent sites without *C. edulis* were also selected to be controls and were situated within 100 m from the occupied sites. *C. edulis* occupied sites where land had been cleared for the construction of electricity power line in the 1960s (pers. comm.) while the control sites had no obvious signs of disturbance.

Three lines of transect measuring 210 to 220 m were placed 60 m each from the disturbed point to the furthest point away from disturbance where *C. edulis* occurred. Following Walker [23] and Gandiwa & Kativu [24]'s set up, 20 x 10 m sampling plots were systematically placed 65 m apart each. Total plots in the sites occupied by *C. edulis* and the control sites were 27 each. Three sections were also demarcated in the sites occupied by *C. edulis* as sections (i) closest, (ii) mid-way, and (iii) furthest from disturbance.

Stem circumferences of each woody species in the study plots were measured at 1.3 m with a tape measure [25] and used to calculate diameter at breast height (circumference/pi). Each stem on multi-stemmed plants was measured and the values summed up to calculate the circumference of the plant [25]. All the woody species in the study plots were also identified *in situ* or at the National Herbarium in Harare.

Differences in species evenness and composition in the control sites and sites occupied by *C. edulis* were verified using the equitability test and Shannon-Weiner diversity index [26]. Separate analyses were done for each site and then averaged. Shapiro–Wilk test was used to test for normality in populations from both the sites occupied by *C. edulis* and the adjacent control sites in SPSS 2007. Significant differences in species richness and evenness in the sites invaded by *C. edulis* and the control sites were assessed using Independent T-test in Microsoft Excel 2007. Significant differences in means of the three sections studied also assessed using One Way ANOVA in Microsoft Excel 2007.

3. Results

Current spatial distribution and location altitude of *C. edulis* in Zimbabwe is as illustrated in **Figure 2**. Manicaland Province recorded the highest number of *C. edulis* presence points, followed by Matebeleland South Province. Of the 161 *C. edulis* presence points recorded, 13% occur at an altitude less than 600 m represented by yellow triangles. 21% occur at an altitude ranging from 600 to 999 m and represented by orange squares. The highest presence points recorded (43%) occur at an altitude between 1000 and 1399 m, and this is followed by 23% which occur at an altitude above 1400 m.

Spatial distribution of *C. edulis* in Zimbabwe and maximum temperature of warmest month at each site are illustrated in **Figure 3**. Sampled *C. edulis* occupies sites with four temperature ranges were recorded out of five. The occupied sites include those with a maximum temperature range of 31–34°C represented by white, 28–31°C represented by yellow, 24–28°C represented by neon/light green and those with the maximum temperature range of 20–24°C represented by forest/dark green.

Spatial distribution and sites' average precipitation of warmest month of *C. edulis* are as illustrated in **Figure 4**. The sampled *C. edulis* is shown to be occupying sites with all five different precipitation ranges. The occupied sites include those with a mean precipitation range of 60 to 300 mm represented by white, 300 to 500 mm represented by yellow, 500 to 800 mm represented by neon-green, 800 to 1000 mm represented by pine-green and those with a mean precipitation range of 1000 to 1261 mm represented by tea-green.

Potential sites of distribution of *C. edulis* in Zimbabwe are illustrated in **Figure 5**. Further *C. edulis* distribution is predicted in the Eastern Highlands (Manicaland Province), parts of Mazowe and Bindura (Mashonaland Central Province) and parts of Matobo (Matebelaland South Province) with a probability of 0.5 to 1 represented by red.

Table 1 shows the percentage contributions of the 19 bioclimatic variables that were used in modelling the potential sites of spread of *C. edulis* in Zimbabwe. Results show that precipitation of driest month (BIO14), mean temperature of warmest quarter (BIO10) and maximum temperature of warmest month (BIO5) had the highest contribution in the modelling of *C. edulis* distribution in Zimbabwe shown in **Figure 4**. Their total contribution was 74.1%.

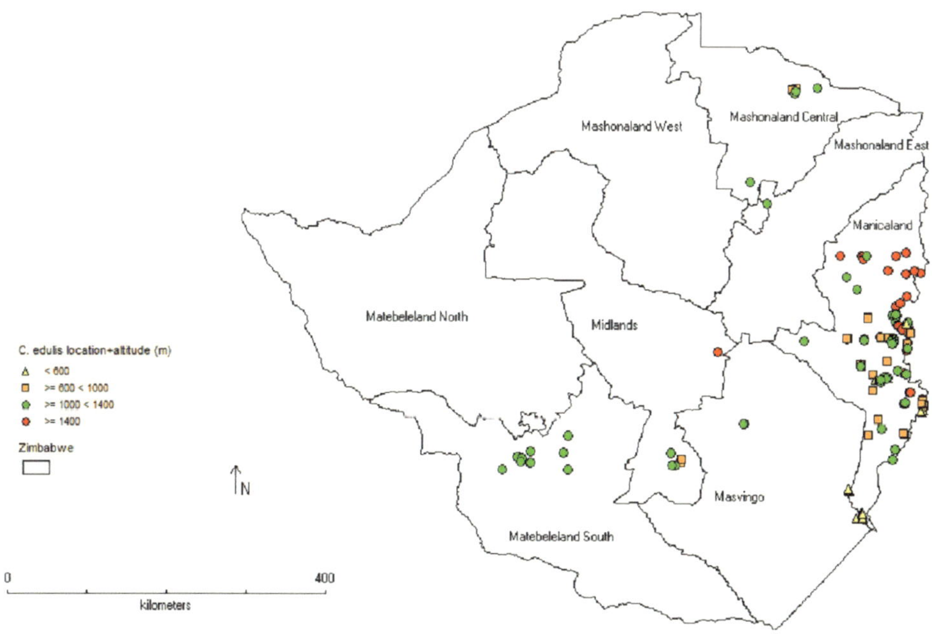

Figure 2.
Distribution and altitudinal location of C. edulis *in Zimbabwe.*

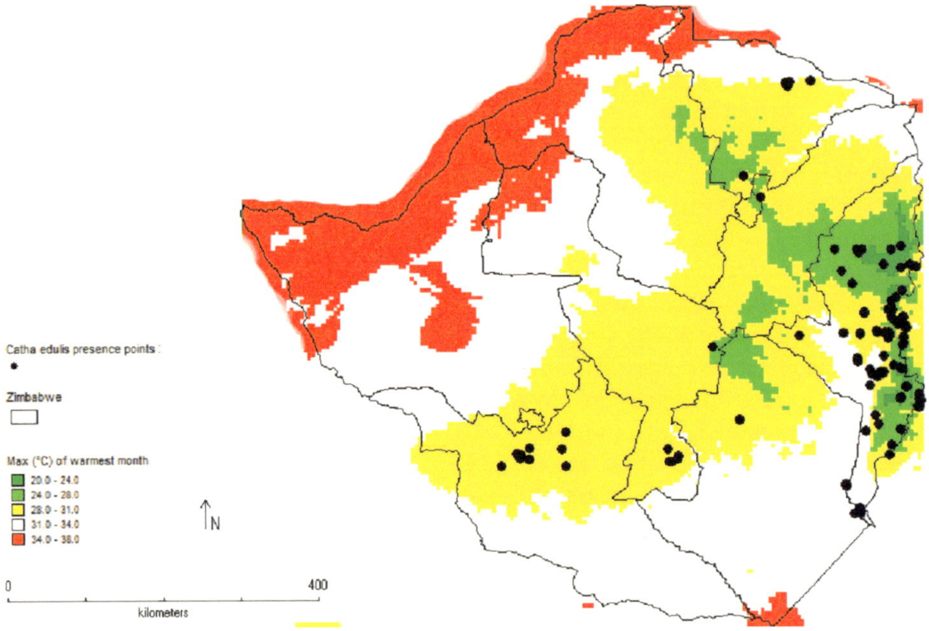

Figure 3.
Distribution of C. edulis *with location maximum temperatures of warmest month in Zimbabwe.*

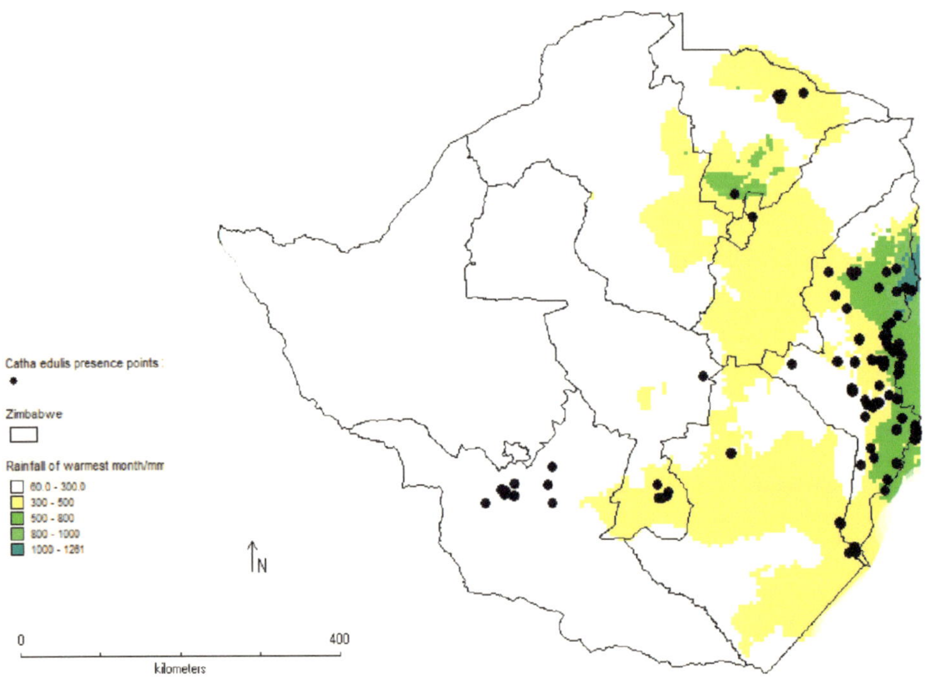

Figure 4.
Distribution of C. edulis *with mean precipitation of warmest month in Zimbabwe.*

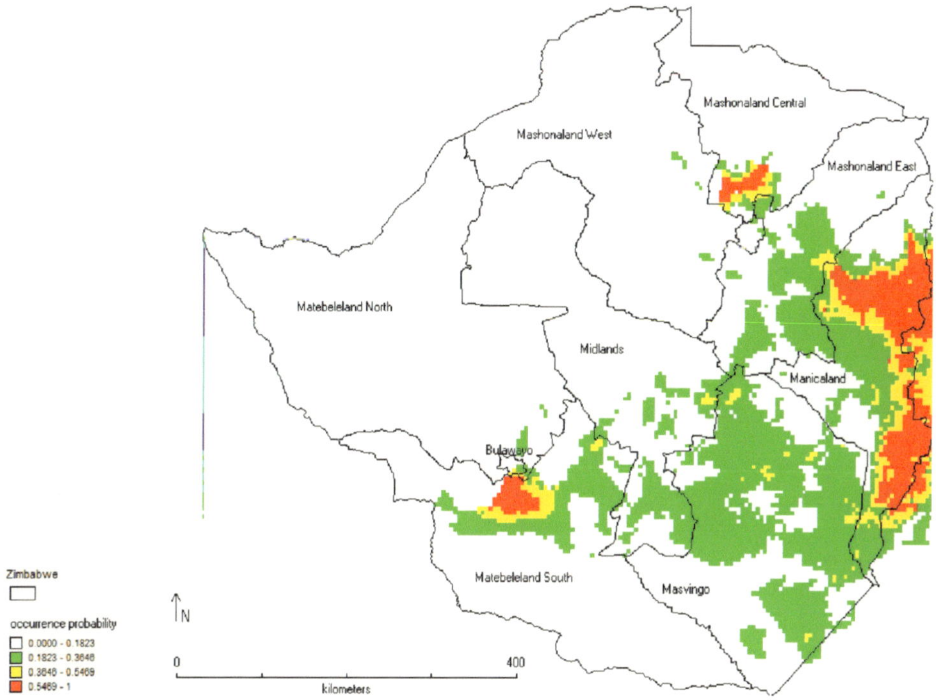

Figure 5.
Potential sites of distribution of C. edulis *in Zimbabwe.*

Figure 6 shows relative importance of the variables with respect to training gain after jackknife test had been done. BIO14 generated the highest gain when used as the only bioclimatic variable of the model and its omission also resulted in the strongest decrease of gain among all the variables. This is consistent with the relevance of bioclim variables test (**Table 1**).

Figure 7 shows the predictive ability of the model in **Figure 5** generated by Maxent using Area Under Curve (AUC) of the Receiver Operating Characteristic (ROC) curve. Random prediction (black line) is a reference line. Both training data (red line) and test data (blue line) shows that the model had a high predictive ability with a high AUC value of 0.95.

Table 2 lists the woody plant species observed in the study plots. Sites invaded by *C. edulis* had 6 species while sites unoccupied by *C. edulis* had 15 species. *Trema orientalis* and *Harungana madagascariensis* were only found in the sites occupied by *C. edulis* while *Bersama abyssinica*, *Heteropyxis dehniae* and *Acacia abyssinica* were recorded in both invaded and control sites.

Sites occupied by *C. edulis* and the adjacent control sites had greater Shapiro–Wilk test values (0.22 and 0.21, respectively) than the alpha value (0.05). The populations are therefore normally distributed. Parametric tests were used to check for differences in species richness and evenness among the populations in the study sites. Means of the sections close to, mid-way and furthest from disturbance were also subjected to parametric tests to check for differences among study sections.

Variable	Name of variable	Percent contribution	Permutation importance
BIO14	Precipitation of driest month	53.7	27.3
BIO10	Mean temperature of warmest quarter	12.6	2.9
BIO5	Maximum temperature of warmest month	7.8	8.4
BIO19	Precipitation of coldest month	6.7	0
BIO16	Precipitation of wettest month	3.1	10.6
BIO7	Temperature annual range	2.6	0
BIO9	Mean temperature of driest quarter	2.3	12.8
BIO4	Temperature seasonality	2.1	3.7
BIO15	Precipitation seasonality	1.4	11.3
BIO18	Precipitation of warmest month	1.3	7.6
BIO3	Isothermality	1.3	4.9
BIO17	Precipitation of driest month	1.2	2.1
BIO13	Precipitation of wettest month	1.2	2.9
BIO8	Mean temperature of wettest month	1.1	0
BIO12	Annual precipitation	1	3.3
BIO6	Mean temperature of coldest month	0.6	0.1
BIO2	Mean diurnal range	0.1	2.2
BIO11	Mean temperature of coldest quarter	0	0
BIO1	Annual mean temperature	0	0

Table 1.
Relevance of each bioclimatic variable for the model.

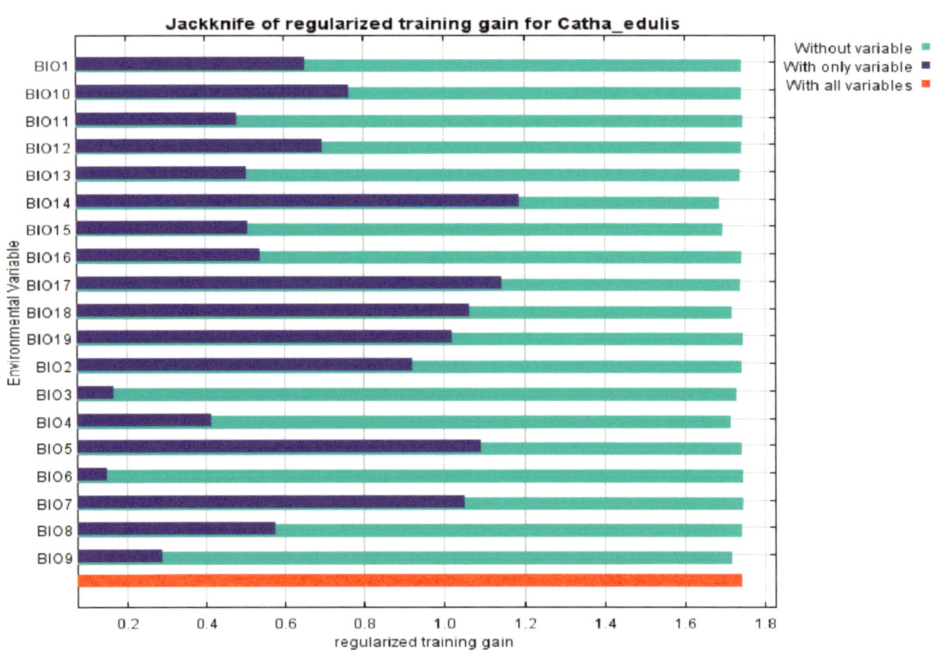

Figure 6.
Results of jackknife evaluation of the relative importance of the variables with respect to training gain.

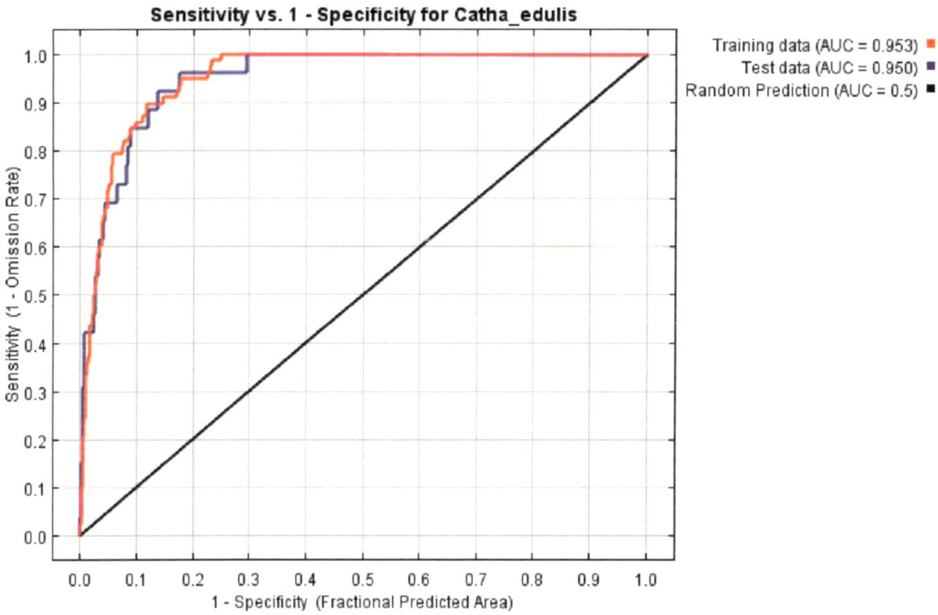

Figure 7.
Evaluation of the predictive ability of the model generated by Maxent using AUC of the ROC curve.

Sites occupied by *C. edulis'* had mean equitability value of 0.11 while the adjacent control sites had a value of 0.45. The latter is closer to 1 while the former is closer to 0. Control sites also had a higher Shannon diversity index of 2.65 compared to that of the sites occupied by the *C. edulis* (0.73). T Stat values in both Equitability test and Shannon diversity index (63.82 and 60.56, respectively) are higher than 2.78 which is the t Critical value. Equitability test and Shannon diversity index's P values of $3.6x10^{-7}$and $4.45x10^{-7}$ are less than 0.05 which is the Alpha value. Therefore, the differences in species evenness and diversity in sites occupied by *C. edulis* and the unoccupied sites is significant.

Results of diameter class distribution of *C. edulis* stems in the section close to, middle section and furthest from disturbance respectively are shown in **Figure 8** and all three sections display an inverse J-distribution.

Percentages of mean number of *C. edulis* stems recorded in the section close to, mid-way and furthest from disturbance in the three diameter classes are summarised in **Table 3**. 0-8 cm diameter class had an average of 502 stems. Section closest to the disturbance had the lowest percentage (20.1%) while the one furthest from the disturbance had the highest percentage (46.4%). The second diameter class (8<16 cm) had an average of 125 stems and 18.4% of these stems were recorded in the section furthest from disturbance while 57.6% were in the section closest to the disturbance. 16<24 cm diameter class had zero recordings in the section furthest from disturbance while 83.3% of the 18 stems recorded in this diameter class were from the section closest to the disturbance.

Results of the One Way Anova show that at least means of *C. edulis* stems in the three diameter classes and the three sections had a significant difference as the 0.05 alpha value was greater than $6.8x10^{-7}$, $1.2x10^{-6}$, and $6.6x10^{-5}$ p values. T-test results showed that the means of *C. edulis* stems in all the three sections were significantly different as 0.05 alpha value was greater than $3.7x10^{-5}$, $1.7x10^{-4}$, $1.4x10^{-3}$, 0.031, $2.31x10^{-4}$, $8.36x10^{-5}$, 0.035, 0.01, 0.005 P values.

Family	Species	*C. edulis* invaded sites	Control sites
Apocynaceae	*Rauvolfia caffra* Sond.		x
Celastraceae	*Catha edulis* Forssk. ex Endl	x	
Clusiaceae	*Harungana madagascariensis* Lam. ex Poir.	x	
Euphorbiaceae	*Macaranga capensis* (Baill.) Benth. ex Sim.		x
Fabaceae	*Acacia abyssinica* Hochst. Ex Beth.	x	x
Fabaceae	*Albizia gummifera* (J.F. Gmel.) C.A. Sm.		x
Fabaceaae	*Newtonia buchananii* (Baker) G.C.C. Gilbert & Boutique.		
Heteropyxidaceae	*Heteropyxis dehniae* Suess.	x	x
Meliaceae	*Ekebergia capensis* Sparrm.		x
Melianthaceae	*Bersama abyssinica* Fresen.	x	x
Moraceae	*Trilepisium madagascariense* DC.		x
Phyllanthaceae	*Bridelia micrantha* (Hochst.) Baill.		x
Proteaceae	*Faurea rubiflora* Marner.		x
Rosaceae	*Prunus Africana* (Hook.f.) Kalkman.		x
Rutaceae	*Calodendrum capense* (L.F.) Thunb.		x
Sapindaceae	*Allophylus abyssinicus* (Hochst.) Radlk.		x
Ulmaceae	*Celtis africana* Burm. f.		x
Ulmaceae	*Trema orientalis* (L.) Blume.	x	

Table 2.
Woody species recorded in sites C. edulis *invaded and non-invaded sites.* **X** *shows that the species is present in the site.*

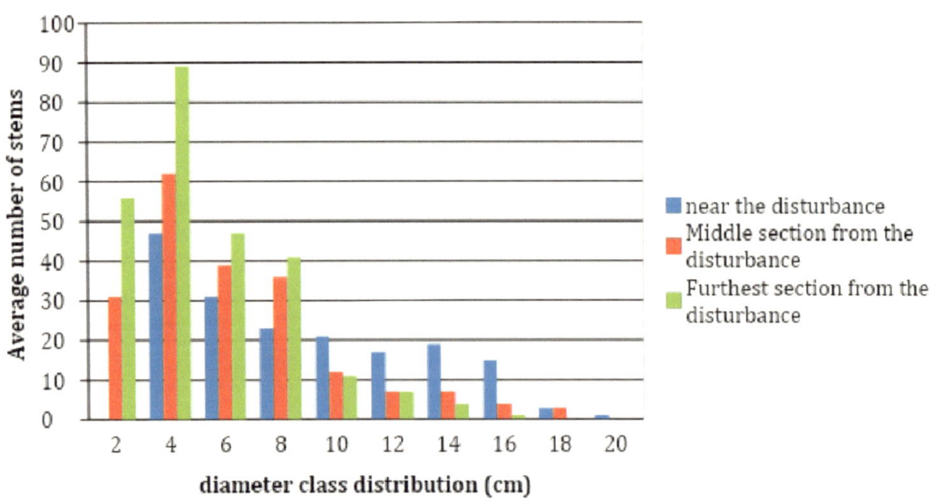

Figure 8.
Diameter class distribution of the mean number of C. edulis *stems observed in sections close to, mid-way and furthest from* C. edulis *invaded sites.*

Diameter class (cm)	near the disturbance	Middle section from the disturbance	Furthest section from the disturbance
0 ≥ 8	(233) 46.40%	(168) 33.50%	(101) 20.10%
8 ≤ 16	(23) 18.40%	(30) 24%	(72) 57.60%
16 ≤ 24	(0) 0%	(3) 16.70%	(15) 83.30%

Table 3.
Percentages of C. edulis *stems recorded in the three diameter classes in study sections.*

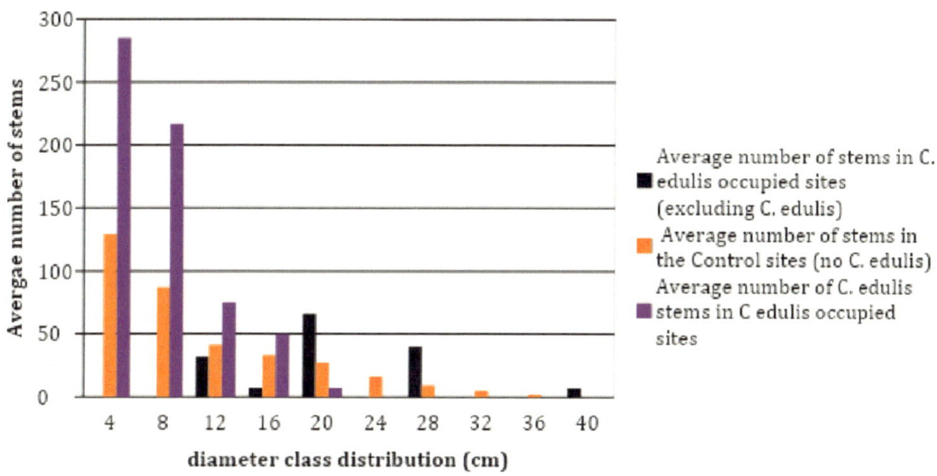

Figure 9.
Diameter class distribution of the (1) mean number of native stems co-occurring with C. edulis*, (2) mean number of native stems invaded sites (3) mean number of* C. edulis *stems in sites invaded sites.*

Results of diameter class distribution are shown in **Figure 9**. An inverse J distribution is indicated for both native plant stems in uninvaded sites and also for *C. edulis* stems in sites it exclusively occupies. An irregular bell shaped distribution is indicated for native stems co-occurring with *C. edulis* with zero stems being recorded in classes 0–8, 21–24 and 29–35 cm.

4. Discussion

Altitude, temperature and precipitation were the primary parameters used in constructing the maps because of their crucial role in determining plant distribution, growth and persistence [27]. Altitude influences such microclimatic conditions as temperature and precipitation [28]. *C. edulis* was found to be predominantly occurring in the Eastern Highlands of Zimbabwe, a region characterised by relatively high altitude, high rainfall and low temperatures. This observation is consistent with what was reported in Yemen and Ethiopia by Al-hebshi & Skaug [29], Zahran et al. [30] and Kandari *et al.* [31]. *C. edulis,* being an evergreen tree [31], is well adapted to these conditions. Broad-leaved subtropical evergreen trees have low water use efficiency. Hence, they thrive in high rainfall conditions. They use the stored, readily available underground water for evaporative cooling during the dry season to avoid excessive light and heat stress which enables them to retain their leaves throughout the dry season [32].

The results, however, show that *C. edulis* also occurs in other parts of Zimbabwe which are at low altitude and characterised by low rainfall and high temperatures. Similar results were summarised by Zahran et al. [30] where different variants of *C. edulis* were distinguished by the Yemenis farmers based on different altitude and climatic conditions under which they thrive. Most plants in dry areas have shallow roots which allow quick uptake of moisture and nutrients near the soil surface [33].

Dessie & Kinlund [34] also reported that *C. edulis* was expanding into relatively higher temperature, lower rainfall and lower altitudinal zones, resulting in the decline of forests in Wondo Genet, Ethiopia. The present study observed similar range expansion trends for *C. dulis* occupied sites in the Eastern Highlands of Zimbabwe. Dessie & Kinlund [34] attributed range expansion of *C. edulis* to various factors, including the species ecological adaptability and its genetic variation [35].

Plants with broader climatic and altitudinal needs are reported to be highly adaptive, with potential to become invasive [36]. Most non-invasive species biological activities and growth occur within specific narrow temperature ranges. Such species require specific amounts of precipitation for physiological processes [37]. However, species with dominance and invasive tendencies tend to tolerate a wide range of climatic and edaphic conditions [36]. In a study to assess invasion risk of plants by Higgins & Richardson [38], a physiologically based species distribution model was used, and it concluded that species that tolerate a wider range of environmental conditions tend to be invasive. Findings of the present study showed that *C. edulis* in Zimbabwe thrives under broad climatic conditions, and has the potential to expand its range of distribution in the Eastern Highlands of the country.

Sites occupied by *C. edulis* had significantly different species evenness and diversity from those not occupied by the *C. edulis*. Other studies involving plant species with invasive properties also showed similar trends [39, 40]. *C. edulis* has a competitive advantage as it produces seeds in large quantities, allowing it to occupy new extended habitats [41, 42].

Plots further away from the point of disturbance had notably younger and higher numbers of *C. edulis* stems in comparison to those closest to the disturbance. *C. edulis* individuals nearest the disturbance were the first ones to occupy the site hence are the oldest. In conducive conditions, the combination of *C. edulis* being tall plus producing relatively smaller seeds [43], allows it to disperse its seeds furthest away from the parent plant [44]. This reduces intra specific competition and enhances increased range distribution for the species [45]. Seeds from tall trees are dispersed from a higher point which is significantly open thereby reduces chances of other plants disturbing them. Shrubs and seedlings of *C. edulis* are therefore more densely populated than the adults [34] in a similar sized area, thus, comprising a higher population density in comparison with a site occupied predominantly by adults.

Fifty percent of the species found in sites disturbed areas are pioneer species [46]. The height, densely packed establishment of *C. edulis* and its closed canopy formation tend to suppress the growth of co-occurring pioneer species. Their seedlings are deprived of light due to the shade formed by *C. edulis'* closed canopy growth form [34]. This strategy has been observed in other species for example, *Impatiens glandulifera* [40]. This species reduces evenness and diversity in habitats it invades by forming a closed canopy over seeds of co-occurring species, thereby depriving them of light which disturbs germination and also cause stunted growth [40].

The observed inverse J-distribution for *C. edulis* implies that the species is in a healthy regenerating state [47]. This distribution pattern is usually displayed when the lower diameter classes have a higher number of individuals which steadily reduce towards the higher diameter classes. Such a distribution ensures sustainability of the population as there will be numerous seeds that can be recruited into the following growth stages [48]. An irregular-bell shaped distribution like the one shown by the co-occuring native species in sites occupied by *C. edulis* suggests that the rate of regeneration is lower that the mortality rate at certain diameter classes [48] therefore the population will be in poor regeneration potential [47]. Hence, their populations are under threat. Native species in sites not occupied by *C. edulis* showed a healthy regeneration potential.

The main driving assumption in population diameter class distribution structures is that a population with more stems in the lower diameter classes compared to the higher ones is constantly regenerating, while that with fewer stems in the lower classes compared to the higher classes is in decline [49]. However, this assumption is not true for all species populations as some have displayed the reverse J-shape distribution but well known to be declining [7]. Some species which do not display the reverse J-shape distribution are actually known to be increasing or their populations are in stable states [50]. These observations have been credited to the different growth rates sometimes found among different size classes. The population diameter distribution method and interpretation, however, has been used successfully in many studies by Souza [47, 48, 51]. While it is not recommended as a sole assessment method, it is a useful basis for management decisions in the absence of complimentary demographic approaches [49].

5. Conclusion

Catha edulis has stable populations with high potential for regeneration in the Eastern Highlands of Zimbabwe, while co-occurring indigenous species are in a decline or unstable state. The persistence of *C. edulis* in sites furthest away from points of disturbance suggests that disturbance is not the only determinant factor for the invasion taking place, which is a worrisome observation for the continued existence of the indigenous species in the area. Future studies must focus on elucidating mechanisms that support and encourage the dominance of *C. edulis* and potentially suppressing co-occurring native species in areas occupied by *C. edulis*.

Acknowledgements

The Norwegian Partnership Programme for Global Academic Cooperation (NORPART) financially supported the study.

Author details

Evelyn Ngarakana[1*], Clemence Zimudzi[1], Shakkie Kativu[1] and Brita Stedje[2]

1 Department of Biological Sciences and Ecology, University of Zimbabwe, Harare, Zimbabwe

2 Natural History Museum, University of Oslo, Oslo, Norway

*Address all correspondence to: evengarakana@gmail.com

References

[1] WWF. 2020. Living Planet Report 2020 – Bending the curve of biodiversity loss. Almond REA, Grooten M, Peterson T. https://www.zsl.org/sites/default/files/LRP%202020%20Full%20report.pdf

[2] Maroyi A. Local plant use and traditional conservation practices in Nhema communal area, Zimbabwe. International Journal of African Renaissance Studies. 2012;7(1):109-128

[3] Tembrock LR, Simmons MP, Richards CM, Reeves PA, Reilley A, Curto MA, et al. Phylogeography of the wild and cultivated stimulant plant qat (Catha edulis, celastraceae) in areas of historical cultivation. American Journal of Botany. 2017;**104**(4):538-549. DOI: 10.3732/ajb.1600437

[4] Wursten B, Ballings P, Palgrave, MC, Hyde M. Catha edulis (Vahl) Forssk.exEndl. 2016. Retrieved July 6, 2017. http://www.zimbabweflora.co.zw/speciesdata/species.php?species_id=137020

[5] Surve I. Chewing the khat is illegal, says Sanab. Independent Online (IOL). 2000. [Retrieved from]: https://www.iol.co.za/news/south-africa/chewing-the-khat-is-illegal-says-san b-35917

[6] Timberlake J, Ballings P, Vidal J Jr, Wursten B, Hyde M, Mapaura A, et al. Mountains of the mist: A first plant checklist for the Bvumba Mountains, Manica Highlands (Zimbabwe-Mozambique). Phyto Keys. 2020;**145**:93-129

[7] Paul A, Khan MF, Das AK. Population structure and regeneration status of rhododendrons in temperate mixed broad-leaved forests of western Arunachal Pradesh, India. Geology Ecology and Landscapes. 2019;**3**(3):168e186. DOI: 10.1080/24749508.2018.1525671

[8] Meekins JF, Mccarthy BC. Effect of population density on the demography of an invasive plant (*Alliariapetiolata*, Brassicaceae) population in a southeastern Ohio Forest. The American Midland Naturalist. 2006;**147**(2):256-278

[9] Dyderski MK, Jagodziński AM. Impact of invasive tree species on natural regeneration species composition, diversity and density. Forests. 2020;**11**:456-476

[10] Ma B, Sun J. Predicting the distribution of Stipapurpurea across the Tibetan Plateau via the MaxEntmodel. BMC Ecology. 2018;**18**(1):1-12. DOI: 10.1186/s12898-018-0165-0

[11] Mugandani R, Wuta M, Makarau A, Chipindu B. Re-classification of agro-ecological regions of Zimbabwe in conformity with climate variability and change. African Crop Science Journal. 2012;**20**(s2):361-369. DOI: 10.4314/acsj.v20i2

[12] Kumirai A, Msimanga A, Munyikwa D, Chidavaenzi R, Muringaniza SJ. Nomination dossier for the proposed Matobo Hills World Heritage. 2003. www.whc.unesco.org/uploads/nominations/306rev.pdf

[13] Ballings P, Wursten B. Vumba – Facts. 2001. https://vumba-nature.com/wp/vumba-nature/vumba-facts/. [Retrieved from]: 22 January 2021

[14] Clark VR, Timberlake JR, Hyde MA, Mapaura A, Coates Palgrave M, et al. A first comprehensive account of floristic diversity and endemism on the Nyanga

Massif, Manica Highlands (Zimbabwe-Mozambique). Kirkia. 2017;**19**:1-53

[15] Cheng D, XU L. Predicting the potential distributions of Senecio vulgaris L. in China. PeerJ Preprints. 2015. DOI: 10.7287/peerj.preprints.1612v1

[16] Perez LV. Principal component analysis to address multicollinearity. 2017. https://www.google.com/url?sa=t&rct=j&q=&esrc=s&source=web&cd=&cad=rja&uact=8&ved=2ahUKEwi75Ijrt-b3AhWoRkEAHflsB9MQFnoECB0QAQ&url=https%3A%2F%2Fwww.whitman.edu%2FDocuments%2FAcademics%2FMathematics%2F2017%2FPerez.pdf&usg=AOvVaw2OOzdrgT--uRZ2WyeooKOZ

[17] Yi YJ, Cheng X, Yang ZF, Zhang SH. Maxentmodeling for predicting the potential distribution of endangered medicinal plant (H. riparia Lour) in Yunnan, China. Ecological Engineering. 2016;**92**:260-269. DOI: 10.1016/j.ecoleng.2016.04.010

[18] Hijmans RJ, Guarino L, Jarvis A, O'Brien R, Mathur P, Cruz CBM, et al. DIVA-GIS Version 5.2 Manual. 2005. Retrieved July 7, 2017, from http://www.diva-gis.org/docs/DIVA-GIS5_manual.pdf

[19] Shabanimofrad M, Yusop MR, Saad MS, Megat Wahab PE, Biabanikhanehkahdani A, Latif MA. Diversity of physic nut (Jatrophacurcasi) in Malaysia: Application of DIVA-geographic information system and cluster analysis. Australian Journal of Crop Science. 2011;**5**(4):361-368

[20] Scheldeman X, Van Zonneveld M. Training Manual on Spatial Analysis of Plant Diversity and Distribution. Analysis. Vol. 3282010. Rome: Biodiversity International; pp. 11-177. DOI: 10.1021/ja0267831

[21] Pearson RG. Species' distribution modeling for conservation educators and practitioners. Lessons in Conservation. 2010;**3**:54-89. DOI: 10.1016/S0140-6736(10)61462-6

[22] Phillips SJ, Dudík M, Schapire R. A maximum entropy approach to species distribution modeling. In Proceedings of the 21st International Conference on Machine Learning. New York: ACM Press. 2004. (pp. 655-662)

[23] Walker BH. An approach to the monitoring of changes in the composition and utilisation of woodland and savanna vegetation. South African Journal of Wildlife Resources. 1976;**6**(1):1-32

[24] Gandiwa E, Kativu S. Influence of fire frequency on *Colophospermummopane* and *Combretumapiculatum* woodland structure and composition in Northen Gonarezhou National Park, Zimbabwe. Koedoe. 2009;**51**(1) Art.#685:13. DOI: 10.4102/koedoe. v51i1.685

[25] Campbell BM, Cunliffe RN, Gambiza J. Vegetation structure and small-scale pattern in miombowoodland, Marondera, Zimbabwe. Bothalia. 1995;**25**(1):121-126

[26] Shannon CE, Weaver W. The Mathematical Theory of Communication. Urbana: University of Illinois Press; 1949

[27] Jin XM, Zhang Y, Schaepman ME, Clevers JGPW, SU Z, Clevers J, et al. Impact of elevation and aspect on the spatial distribution of vegetation in the Qilian and mountain area with remote sensing data. The International Archives of the Photogrammetry, Remote Sensing and Spatial Information Sciences. 2006;**37**:1385-1390

[28] He X, Hou E, Liu Y, Wen D. Altitudinal patterns and controls of

plant and soil nutrient concentrations and stoichiometry in subtropical China. Scientific Reports. 2016;**6**(24261):1-9. DOI: 10.1038/srep24261

[29] Al-Hebshi NN, Skaug N. Khat (Catha edulis)— an updated review. Addiction Biology. 2005;**10**(December):299-307. DOI: 10.1080/13556210500353020

[30] Zahran MA, Khedr A, Dahmash A, El-Ameir YA. Qat farms in Yemen: Ecology, dangerous impacts and future promise. Egyptian Journal of Basic and Applied Sciences. 2013;**1**(1):1-8. DOI: 10.1016/j.ejbas.2013.09.002

[31] Kandari LS, Yadav HR, Thakur AK, Kandari T. Chat (Catha edulis): A socio economic crop in Harar Region, Eastern Ethiopia. SpringerPlus. 2014;**3**(579):1-9. DOI: 10.1186/2193-1801-3-579

[32] Tomlinson KW, Poorter L, Sterck FJ, Borghetti F, Ward D, De Bie S, et al. Leaf adaptations of evergreen and deciduous trees of semi-arid and humid savannas on three continents. Journal of Ecology. 2013;**101**(2):430-440. DOI: 10.1111/1365-2745.12056

[33] Baudena M, Dekker SC, Van Bodegom PM, Cuesta B, Higgins SI, Lehsten V, et al. Forests, savannas, and grasslands: Bridging the knowledge gap between ecology and dynamic global vegetation models. Biogeosciences. 2015;**12**(6):1833-1848. DOI: 10.5194/bg-12-1833-2015

[34] Dessie G, Kinlund P. Khat expansion and forest decline in Wondo Genet, Ethiopia. Geografiska Annaler: Series B, Human Geography. 2008;**90**(2):187-203. DOI: 10.1111/j.1468-0467.2008.00286.x

[35] Al-Thobhani MAH, Sathyanarayana NB, Simon L, Sondur S. First comparative genotypic study on Khat(CathaedulisForsk.) genotypes from Yemen. Middle Eastern and Russian Journal of Plant Science and Biotechnology. 2008;2(1), 1-8. [Retrieved from]: http://www.globalsciencebooks.info/Online/GSBOnline/images/0812/MERJPSB_2 (1&2)/MERJPSB_2(1)1-8o.pdf

[36] Ratnayake RMCS. Why plant species become invasive? Characters related to successful biological invasion. In: National Symposium on Invasive Alien Species. 2014. (pp. 22-42). [Retrieved from]: https://www.researchgate.net/publication/273452847_Why_plant_species_become_invasive

[37] Llewellyn LM. Range plant growth and development are affected by climatic factors. 2013. [Retrieved from]: https://www.ag.ndsu.edu/DickinsonREC/grazing-handbook-files/topic1-volume1-ch2-report1.pdf

[38] Higgins SI, Richardson DM. Invasive plants have broader physiological niches. Proceedings of the National Academy of Sciences of the United States of America. 2014;**111**(29):10610-10614. DOI: 10.1073/pnas.1406075111

[39] Hejda M, Pysek P, Jarosík V. Impact of invasive plants on the species richness, diversity and composition of invaded communities. Journal of Ecology. 2009;**97**:393-403

[40] Hulme PE, Bremner ET. Assessing the impact of Impatiens gland ulifera on riparian habitats: Partitioning diversity components following species removal. Journal of Applied Ecology. 2006;**43**:43-50

[41] Landenberger RE, Kota NL, McGraw JB. Seed dispersal of the non-native invasive tree *Ailanthus altissima* into contrasting environments. Plant Ecology. 2007;**192**(1):55-70

[42] Wan J, Wang C, Yu F. Wind effects on habitat distributions of wind-dispersed invasive plants across different biomes on a global scale: Assessment using six species. Ecological Informatics. 2017;**42**:38-45

[43] van der Vossen, H.A.M., 2002. Catha edulis (Vahl) Forssk.exEndl. [Internet] Record from PROTA4U. Oyen, L.P.A. & Lemmens, R.H.M.J. (Editors). PROTA (Plant Resources of Tropical Africa/Ressourcesvégétales de l'Afriquetropicale), Wageningen, Netherlands. https://uses.plantnet-project.org/en/Catha_edulis_(PROTA). Accessed 3 September 2018.

[44] Thomson F, Moles A, Auld TD, Kingsford RT. Seed dispersal distance is more strongly correlated with plant height than with seed mass. Journal of Ecology. 2011;**99**(6):1299-1307

[45] Yang X, Zhang W, He Q. Effects of intraspecific competition on growth, architecture and biomass allocation of Quercus Liaotungensis. Journal of Plant Interactions. 2019;**14**(1):284-294. DOI: 10.1080/17429145.2019.1629656

[46] Hyde MA, Wursten BT, Ballings P, Coates Palgrave M. Flora of Zimbabwe: Home page.2020. https://www.zimbabweflora.co.zw/index.php, retrieved 24 December 2020

[47] Mekonen T, Ayele B, Ashagrie Y. Woody plant species diversity, structure and regeneration status of Woynwuha Natural Forest, North West Ethiopia. Asian Journal of Ethnopharmacology and Medicinal Foods. 2015;**01**(01):07-19

[48] Nabasumba D, Eilu G, Bahati J, Katwesigye G. Utilisation and conservation status of indigenous woody plant species in a sedentary pastoral production system in south-western Uganda. Uganda Journal of Agricultural Sciences. 2016;**17**(2):125-137

[49] Virillo CB, Martins FR, Tamashiro JY, dos Santos FAM. Is size structure a good measure of future trends of plantpopulations? An empirical approach using five woody species from the Cerrado (Brazilian savanna). ActaBotanicaBrasilica. 2011;**25**(3):593-600

[50] Condit R, Sukumar R, Hubbell SP, Foster RB. Predicting population trends from size distributions: A direct test in a tropical tree community. The American Naturalist. 1998;**152**:496-509

[51] Souza AF. Ecological interpretation of multiple population size structures in trees: The case of Araucaria angustifolia in South America. Austral Ecology. 2007;**32**:524-533. DOI: 10.1111/j.1442-9993.2007.01724.x

Chapter 5

The Rising Threat of Invasive Alien Plant Species in Agriculture

Melekote Nagabhushan Arun, Rapolu Mahender Kumar, Banugu Sreedevi, Guntupalli Padmavathi, Pallakonda Revathi, Neha Pathak, Dayyala Srinivas and Boya Venkatanna

Abstract

A species is considered to be invasive if it establishes, persists, and spreads widely inside a natural ecosystem, stunting the growth of native plants and giving them room to overtake crops and native plants. Non-native plant species that have been brought into a new geographic area and have a negative effect on the ecosystems supporting horticulture and agriculture are known as invasive plant species. Invasive/noxious weeds, which are widely distributed in many types of ecosystems, significantly reduce crop production. Compared to native species, invading plant species have a higher potential to move their niche more rapidly and are more likely to adapt to new environments. The timing, speed, and longevity of seed germination have indeed been discovered to change as a result of climate change, which has consequences for plant invasions. More than native plant species, invasive plant species gain from atmospheric carbon dioxide (CO_2) enrichment, greenhouse gas emissions, and global warming. A loss of native biodiversity due to invasive species includes species extinction, changes in hydrology, and altered ecosystem function.

Keywords: invasive alien plant, global warming, climate change, weed shift, crop weed competition

1. Introduction

The invasive species is significant on a global importance. A non-native plant or other organism is considered an invasive species if it completely takes over an ecosystem and damages both its structure and function. Invasive species displace or harm local wildlife and plants, frequently posing major challenges to the area's biodiversity and creating unfavorable environmental conditions. There are no geographical limitations to the type or spread of invasive species. The greatest direct economic losses in crop production are caused by invasive weeds. One of the major direct causes of environmental change on a worldwide scale with a large ecological impact is biological invasion. The potential impact of invasive alien plant species on global agriculture, which continues to affect food security globally, could be significant [1]. The economic cost of plant invasion to agriculture is growing due to the increasing number of new introductions

IntechOpen

which create a tremendous impact on crop production. The invasive alien plants / weeds have many similar biological attributes/traits relating to high reproduction and stress tolerance. The traits include germination of seeds, rapid seedling growth, vegetative and sexual reproduction at early stage aggressive spread by runners or rhizomes, diverse dispersal mechanisms and the ability to tolerate a wide range of environmental condition.

Warming of the earth surface is inevitable due to influence of greenhouse gas emission and instinctive climate variability. The average temperature of the earth has increased considerably by 1.53° C from 1900 to 2020 which has impacted the growing seasons of crops leading to reductions in crop yields [2]. Ramification of crop productivity is considerably noticeable on crop productivity. Potential growth and distribution of invasive plant species are accelerated by climate changes like rise in temperature, atmospheric carbon-dioxide level, nitrous oxide, methane gas emission, extreme weather conditions and change in rainfall pattern. Invasive plants reduce agricultural productivity by way of considerable mechanisms: competition for light, water, nutrient, allelopathy effects and decrease the crop yields and inhibition of seed germination [3].

Invasive and climate change are two of the primary factors which alter ecological systems. Temperature, precipitation, nitrogen, carbon dioxide, and measurements of organismal response in field conditions are manipulations of factors anticipated to vary with climate change. Therefore, the objective of the book chapter is to discuss the effect of climate change on invasive weed floral composition, distribution and effect on crop production.

2. Influence of invasive alien plants on N and P Pool in soil and plant

The success of invasion is mainly the result of the status of soil or growing environment of invasive alien plants (**Figure 1**). NH_4^+ concentration in soil evaded by *Chromolena odorata* was 1.43 times higher than native soil [4] and the NH_4^+ concentration of the soil invaded by *Ageratina adenophora* was 1.56–2.10 times that of soil with native plant species. The differences in soil properties and functioning point towards the contribution of root exudes and higher productivity of litter and their associated spatial variability [5]. Invasive alien plants have advantages over native plants which include higher photosynthetic rate, speeder growth rate, larger reproductive output, larger biomass, lower carbon-to-nutrient ratio in tissue, stronger capacity for nutrient absorption and higher plasticity levels [6]. The invasive species exhibit more strategic advantage for nutrient use over native plants [7] and hence lead to a greater enhancement in the N and P mineralization rates of the soil [8].

Invasive alien weeds such as *Bidens pilosa*, *Microstegium vemineum* and *Mikania micrantha* absorb nitrate over ammonium which causes competition with native crops in nitrate rich soils [9]. African native weed *Andropogon gayanus* was found to directly alter soil structure in tropical Australia which was attributed to the weed accelerating the ammonia process and increasing soil ammonium availability to four timed that of native plant soil, with more than six times higher uptake rate of ammonium than native species. The availability of N, P and N/P ratio profoundly impacted interspecific competition between invaded habitat and native weeds. Hence nutrient deposition promoted the invasiveness of alien plants in the ecosystem [10].

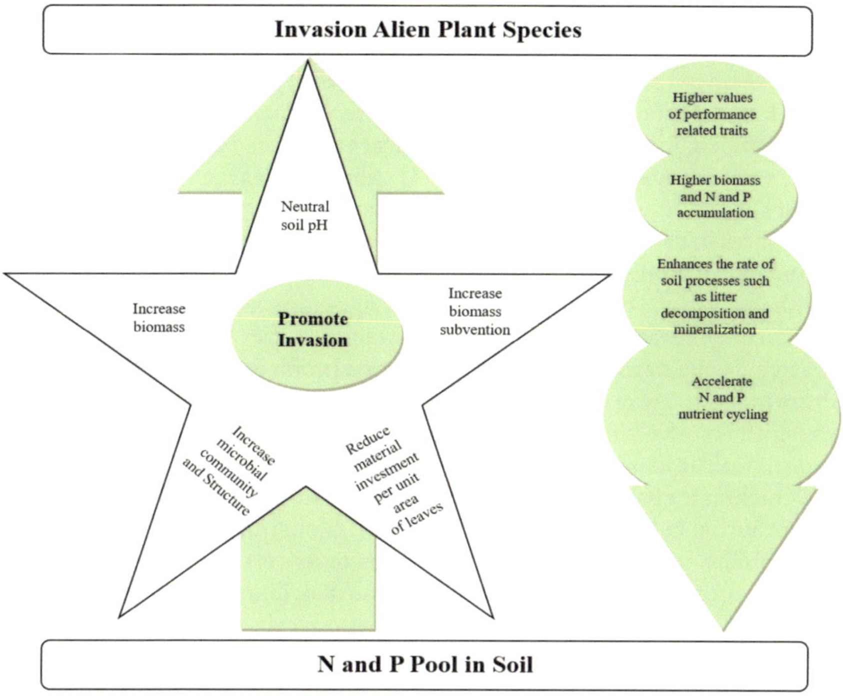

Figure 1.
Influence of nutrient fluctuation caused by N and P on the invasiveness of alien plants.

3. Climate change and weed invasion

Climate is known as the main environmental driver of species distribution, and there have been extensive studies on the distribution areas of invasive plants in determining invasive spreading [11]. Movement of weed species from native range to new areas naturally makes non-native species invasive with negative impact on native species of arable ecosystem. Climate change provides the opportunity for weeds to invade new ecosystems. Climate changes enhances the adaptability of the introduced plants to the new host range and increasing the risk of invasion in native and managed ecosystem since they are suited to new environments and successful in resource utilization in elevated CO_2 concentration. Interactions between climate change and management practices may turn invasive species with high potential to spread widely causing impact on productivity. Weeds can be highly response to increased CO_2 concentration [12]. Invasive and climate change are two of the primary factors which alter ecological systems.

Manipulation of factors likely to change with climate is temperature, precipitation, nitrous oxide levels and carbon dioxide and measurement of organismal response under field conditions integrate the biotic and abiotic factors individuals. Invasive species are most commonly defined as a non-native plant or other organism that dominates the encountered ecosystem and impairs its function and structure. Invasive species displace or damage native fauna and flora often posing serious threats to local biodiversity and causing adverse environment stress. Invasive alien species are one of the major threats to global and local diversity. The threats caused by invasive plant species in agricultural ecosystem include hybridization and species completion.

Global warming may result in the expansion in the habitat range of invasive species and the contraction or displacement of the habitat range of indigenous species [13].

Plant invasion is a serious threat to global biodiversity and hence deleterious to ecology and nature biodiversity. Invasive plants metamorphose the landscape ecology in a highly complex manner leading to ecological explosion. Global terrestrial crops are invaded by various invasive weed species [14]. Alien species that endanger ecosystems, habitats, or species, as well as agricultural production, are considered invasive species. Recent advances in genetics and molecular biology have paved the way for impacts on ecology and global biodiversity. The histories of invasion and agriculture are internally linked with many crops being invasive species. Agricultural biotechnology which is the insertion of genes into crops has generated concern over the risk of producing new invasive species or exacerbating current weed problems. The modern intensive agriculture paved the way for invasive weeds to spread across the globe. Land use changes which is conversion of forests/grasslands into agroecosystem habitat fragmentation as well as increase the level of organic pollutants resulting in the increase level of CO_2/climate change. Global climate changes are directly linked to biological invasions resulting in biodiversity loss (**Figure 2**). Global change stressors like climate change variability and changes in land use are major drivers of ecosystem alterations. Climate is the principal determinant of vegetation distribution from regional to global levels. The global climate is changing; along with measuring temperature and CO_2 level changes are considered major drivers of climate change [15].

Climate conditions exert a significant influence in the spread, population dynamics, life cycle duration, infestation pressure and the overall occurrence of invasive species. Invasive weeds will be influenced by climate change. The direct and indirect consequence of increasing CO_2 or climate change which differentially affects the growth of invasive weeds and crops will alter crop weed competitive interactions. Climate change has a big impact on invasive weed species' distribution, population dynamics, life cycles, pressure from infestations, and overall occurrence [16]. *Parthenium hysterophorus* L. is an invasive weed species worldwide. It is considered as one of the worst weeds in the world due to high fecundity that is ability to produce

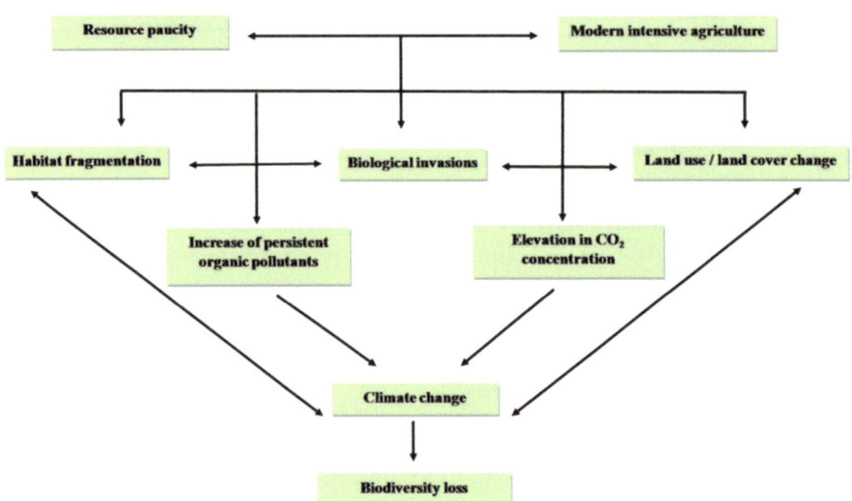

Figure 2.
Paradigm of global inter connected ecological and invasive plant issues.

lot of seeds (20,000 seeds/plants, rapid germination, fast growth rate and threat to crops. The seeds can germinate in a wide range of temperature and cause 40–97 percent yield reduction in crops [17].

3.1 Consequences due to invasive weeds

Biological invasion has become one of the major causes of economic and environmental damage in most of the countries across the world and its impact have been predicted to increase ever further under future climatic conditions. The Convention on Biological Diversity (1992) emphasized biological invasion as one of the drivers of biodiversity decline. Invasive potential of species enables weeds to be successful invaders and colonizers of the novel environments whether introduced deliberately or accidently. Developing regions are fast witnessing the change across all countries. Losses caused by invasive weeds are thrashing of biodiversity from native ecosystems [18], alteration in ecosystem, decline in abundance and richness of native flora and alteration in community structure. The risk of introduction of alien invasive weeds has enhanced due to global climate change. It is estimated 20–30 percent of all introduced species worldwide cause a problem. The impact of climate change on invasive weeds indicated that weeds on the whole have a large growth in the increase in atmosphere CO_2 concentration relation to plant species and rising CO_2 can be sustainable for invasive noxious species within plant communities [19]. Global efforts are very important to control the invasive weed species. Differences between native and exotic plant species in their mode of resource utilization may cause a change in soil structure, its profile, decomposition, nutrient content of soil and moisture availability. Invasive weed species is a serious hindrance to conservation and sustainable use of biodiversity. The impact of climate change on invasive weeds and indicated that the invasive, noxious weeds on the whole have a larger growth in the projected increases in atmospheric CO_2 concentration in relation to other plant species [20]. Ecological integrity and biodiversity of agriculture ecosystems have been seriously threatened by expansion of invasive weed species across globe. Climate change induced transformations in the invasive weed flora of arable ecosystems. Thermophilic weeds and late emerging invasive weeds have become more abundant in cropping system. Prominent invasive weed species like *Lantana camara*, *Mikania micrantha*, *Chromolaena odorata*, *Eupatorium adenophorum*, *Cytisus scoparius*, *Mimosa invisa*, *Parthenium*, *hysterophorus* among terrestrial exotics and *Eichhornia crassipes* and *Pistia stratiotis* among aquatic have posed greater threat to the native crop flora [21].

3.2 Invasive alien species/weeds and their distribution

Invasive alien weed species shift is an important aftermath of global climate change in ecosystem that affects weed management strategies and agricultural productivity (**Table 1**). Climate change is viewed as a cause in accelerating the rate of invasion by alien species in addition to the globalization of anthropogenic activities. *Rottboellia cochinchinensis* is an aggressive invasive species native to Asia. The species is known worldwide for invading crops and disturbed habitats in tropical and subtropical regions. The species spread from South America to Asia in 1961 through the seeds accidentally mixed with rice seeds and found to contain in 27 countries [22]. *Imperata cylindrica* (L.) P. Beauv, is one of the ten worst weeds a perennial grass native to South East Asia is a wide spread invader to warmer regions [23, 24]. *Persicaria glabra* is an invasive weed plant native to North America and Eurasia belonging to the family Polygonaceae

Scientific Name	Family	**Origin**	**Distribution**	**Propagation**
Acanthospermum hispidum	Asteraceae	Brazil (South America)	Widespread in the tropics.	Seed
Ageratina adenophora (Spreng.) King & Robinson	Asteraceae	Mexico (Central America)	Tropical and Subtropical region	Seed
Ageratum conyzoides	Asteraceae	Tropical America	Tropical and Subtropical region	Seed
Alternanthera paronychioides A. St.Hil	Amaranthaceae	Colombia (Tropical America)	Asia and Africa	Seed
Alternanthera philoxeroides	Amaranthaceae	South America	China, Australia, Thailand	Vegetative
Alternanthera pungens	Amaranthaceae	Tropical America	Tropical Africa, Asia, and Australia	Seeds, Vegetative
Alternanthera tenella	Amaranthaceae	Tropical America	Tropical Africa and Asia	Seed
Ambrosia artimisiifolia	Asteraceae	North and Central America	Europe, Africa and Asia	Seed
Ambrosiat trifida	Asteraceae	North America	Temperate Europe and Asia	Seed
Ammania baccifera	Lythraceae	Tropical Africa	Tropical Asia, Africa and America	Seed
Argemone Mexicana	Papaveraceae	Tropical and South America	Tropical and Subtropical region	Seed
Asteracantha lonfifolia	Acanthaceae	Tropical Asia	Tropical Africa and America	Seed
Bidens pilosa	Asteraceae	Tropical America	Tropical of regions Africa and Asia	Seed
Blumea eriantha DC.	Asteraceae	Tropical America	Asia and Africa	Seed
Blumea lacera (Burm. f.) DC.	Asteraceae	Tropical America	Asia, tropical Africa and Australia	Seed
Capsella bursa	Brassicaceae	Mediterranean Region	Wide temperate region	Seed
Cassia rotundifolia Pers.	Caesalpiniaceae	Tropical South America	Tropical and Subtropical Africa Asia	Seed
Celosia argentea	Amaranthaceae	Tropical Africa	Tropical and Sub tropical Asia	Seed
Centella asiatica	Apiaceae	Tropical Asia	Widespread in the tropical regions	Seed, Vegetative
Chenopodium album	Amaranthaceae	Europe	Temperate and Subtropical region	Seed
Chloris barbata Sw.	Poaceae	Tropical America	Tropical and Sub tropical Asia	Seed
Chromolaena odorata	Asteraceae	Tropical America	Humid tropical Asia and Africa	Seed

Scientific Name	Family	**Origin**	**Distribution**	**Propagation**
Cirsium arvense	Asteraceae	South eastern Europe	Subtropical and temperate region	Seed, Vegetative
Cleome gynandra L.	Cleomaceae	Tropical America	Tropical and Subtropical worldwide	Seed
Cleome rutidosperma DC.	Cleomaceae	Tropical America	Tropical Africa Asia and Australia	Seed
Cleome viscosa L.	Cleomaceae	Tropical America	Tropical and Subtropical region	Seed
Commelina benghalensis	Commelinaceae	Tropical Asia	Tropical Africa and Subtropical Asia	Seed, Vegetative
Cuscuta chinensis	Cuscutaceae	Mediterranean	Distributed worldwide	Seed
Cuscuta reflexa	Cuscutaceae	Tropical Asia	Distributed worldwide	Seed
Cyanotisaxillaris	Commelinaceae	Indian sub-continent	South East Asia and Australia	Seed, Vegetative
Cyperus difformis	Cyperaceae	Tropical America	Distributed worldwide	Seed
Cyperus iria	Cyperaceae	Tropical America	Distributed worldwide	Seed
Cytisus scoporius	Fabaceae	Central and Southern Europe	Temperate and sub-tropical region	Seed
Dactylactenum aegyptium	Poaceae	Tropical Africa	Tropical, Subtropical and warm temperate	Seed
Datura innoxia	Solanaceae	Tropical America	Tropical and Subtropical Asia & Africa	Seed
Datura metel	Solanaceae	Tropical America	Tropics and Subtropics worldwide	Seed
Digera muricata (L.) Mart.	Amaranthaceae	Southwest Asia	Tropical Africa and Malesia	Seed
Digitaria sanguinalis	Poaceae	Eurasia	Temperate warm region of world	Seed
Dinebra retroflexa (Vahl) Panz.	Poaceae	Tropical America	Through tropical and South Africa	Seed
Echinochloa colona	Poaceae	Tropical South America	Worldwide Tropics and Subtropics	Seed
Echinochloa crusgalli	Poaceae	Tropical South America	Worldwide Tropics and Subtropics	Seed
Eclipta prostrata	Asteraceae	Tropical America	Tropical, Subtropical and warm temperate	Seed
Eichhornia crassipes	Pontederiaceae	Tropical America	Distributed worldwide	Vegetative
Eleusine indica	Poaceae	Eurasia	Distributed worldwide	Seed
Elytrigia repens	Poaceae	Europe	Distributed to temperate region	Seed

Scientific Name	Family	**Origin**	**Distribution**	**Propagation**
Equisetum arvense	Equisetaceae	Europe	Distributed Europe and Asia	Seeds and Rhizomes
Euphorbia cyathophora Murray	Euphorbiaceae	South America	Subtropical areas worldwide	Seed
Euphorbia hirta	Euphorbiaceae	Tropical America	Widespread Tropical and Subtropical	Seed
Evolvulus nummularius (L.)	Convolvulaceae	South America	Tropical and Subtropical regions	Seed
Fimbristyllus dichotoma	Cyperaceae	Tropical America	Distributed worldwide	Seed
Flaveria trinervia (Spreng.) C. Mohr.	Asteraceae	Tropical Central America	Tropical regions	Seed
Glechoma hederacea	Laminaceae	Europe	North America, Australia and New Zealand	Vegetative and Seed
Gnaphalium pensylvanicum Willd.	Asteraceae	Tropical America	Distributed worldwide	Seed
Gnaphalium polycaulon Pers.	Asteraceae	Tropical America	South America, Tropical Asia and Africa	Seed
Gomphrena serrata L.	Amaranthaceae	Tropical America	Distributed worldwide	Seed
Impatiens capensis	Balsiminaceae	North America	Temperate region	Seed
Imperata cylindrica	Poaceae	Tropical America	Tropical and Warm Temperate region	Seeds
Ipomoea carnea	Convolvulaceae	South America	Tropical and Subtropical region.	Seed
Kyllinga nemorallis	Cyperaceae	South East Asia	Distributed worldwide	Seedsand Rhizomes
Lagascea mollis Cav.	Asteraceae	Tropical Central America	Tropical and Subtropical regions	Seed
Lantana camara	Verbenaceae	Tropical America	Tropical and Subtropical regions	Seed
Leersia oryzoides	Poaceae	Central America	Tropical and Subtropical regions	Seed
Leontodon taraxacum	Asteraceae	Europe	Distributed worldwide temperate region	Seed
Leptochloa chinensis (L.)	Poaceae	Tropical Asia	Africa, Central and South America	Seed and Vegetative
Leptochloa uninervia (J. Presl) Hitchc. & Chase	Poaceae	Central America	Distributed worldwide	Seed and Vegetative
Ludwigia adscendens (L.) Hara	Onagraceae	Tropical America	South East Asia and Malesia	Seed

Scientific Name	Family	Origin	Distribution	Propagation
Ludwigia octovalvis (Jacq.) Raven	Onagraceae	Tropical Africa	Throughout the Tropical world	Seed
Ludwigia perennis	Onagraceae	Tropical Africa	Throughout the Tropical world	Seed
Marselia quadrifolia	Marsileaceae	Southern and Central Europe	North America and Asia	Rhizomes
Merremia aegyptia (L.) Urban.	Convolvulaceae	Tropical America	Worldwide Tropical and Subtropical	Seed
Mikania micrantha Kunth	Asteraceae	Tropical America	Tropical area Africa and Asia	Seed
Mimosa pudica	Mimosaceae	South and Central America	Tropical regions of the World	Seed
Mimosa invisa	Mimosaceae	South and Central America	Tropical regions of the World	Seed
Mirabilis jalapa L.	Nyctaginaceae	Peru	Warmer parts across World	Seed
Monochoria vaginalis (Burm.f.) C. Presl.	Pontederiaceae	Tropical America	Tropical and Subtropical wet areas	Seed
Nastridium indicum	Tropaeolaceae	South America	Distributed worldwide	Seed
Nicotiana plumbaginifolia Viv	Solanaceae	Tropical America	Tropical regions of the World	Seed
Panicum repens	Poaceae	Africa	Tropics and Subtropics	Seed and rhizomes
Parthenium hysterophorus	Asteraceae	Tropical and North America	Throughout the World	Seed
Papspalum dilatum	Poaceae	South America	Humid Tropics and Subtropics	Seed
Paspalaum distichum	Poaceae	Tropical and Subtropical America	Tropical and Subtropical region	Seed
Paspalum hydrophyllum	Poaceae	South America	Tropical Asia, Africa and Australia	Seed
Passiflora foetida L.	Passifloraceae	Tropical and South America	Tropical region of Asia and Africa	Seed
Pennisetum purpureum	Poaceae	Tropical America	Tropical and Subtropical region	Seed
Phyla nodiflora	Verbenaceae	South America	Tropical and Subtropical region	Seed
Phyllanthus tenellus	Euphorbiaceae	Mascarene Islands	Africa, Southern Europe and Asia	Seed
Physalis angulata L.	Solanaceae	Tropical America	Asia and Africa	Seed
Pistia stratiotes L.	Araceae	Tropical America	Tropical and Subtropical region	Vegetative

Scientific Name	Family	Origin	Distribution	Propagation
Plantigo lanceolata	Plantaginaceae	Eurasia	South Asia, Australia and North America	Seed
Portulaca oleracea	Portulacaceae	Tropical Central America	Tropical and Subtropical region	Seeds
Portulaca quadrifida	Portulacaceae	Tropical South America	Africa and Tropical Asia	Seed
Prosopis juliflora (Sw.) DC.	Mimosaceae	Mexico	Tropical and Subtropical region	Seed
Rotala densiflora	Lythraceae	Tropical Asia	Tropical Africa America and Australia	spores
Ruellia tuberosa L.	Acanthaceae	Tropical America	South East Asia and Tropical Africa	Seed
Salvinia molesta	Salviniaceae	South Eastern Brazil	Wide spread across tropical world	Vegetative
Sida acuta Burm.f.	Malvaceae	Tropical America	Pacific and South East Asia	Seed
Solanum seaforthianum Andre	Solanaceae	Brazil	Worldwide distribution	Seed
Solanum viarum Dunal	Solanaceae	Tropical America	Tropical and Subtropical region	Seed
Sonchus oleraceus L.	Asteraceae	Mediterranean	Tropical and Subtropical region	Seed
Sonchus asper Hill	Asteraceae	Mediterranean	Tropical and Subtropical region	Seed
Stylosanthes hamata (L.) Taub.	Papilionaceae	Tropical America	Tropical Africa and Asia	Seed
Stachytarpheta jamaicensis (L.) Vahl	Verbenaceae	Tropical America	Subtropical Asia Africa and Oceania	Seed
Stachytarpheta urticaefolia (Salisb.) Sims	Verbenaceae	Tropical America	Tropical Africa, Asia and Pacific region	Seed
Stellaria media	Caryophyllaceae	Europe	Throughout the world	Seed
Synadenium grantii Hook. F.	Euphorbiaceae	Tropical Africa	Tropical region of America and Asia	Seed
Synedrella nodiflora (L.) Gaertn.	Asteraceae	West Indies	Warmer region of the world	Seed
Taraxacum officinale	Asteraceae	Europe	Temperate region of the world	Seed
Tribulus terrestris	Zygophyllaceae	Tropical America	Warm Temperate region of Eurasia, Africa	Seed
Tridax procumbens	Asteraceae	Tropical Central America	Warm Temperate and Tropical region	Seed
Turnera subulata J.E. Smith	Turneraceae	Tropical America	Tropical region of Asia and Africa	Seed

Scientific Name	Family	Origin	Distribution	Propagation
Turnera ulmifolia L.	Turneraceae	Tropical America	Africa, South East Asia and Tropical Island	Seed
Typha angustata	Typhaceae	Tropical America	Asia, North Africa and South Europe	Seed
Ulex europaeus L.	Papilionaceae	Western Europe	Tropical Africa and Asia and Australia, NZ	Seed
Urena lobata L.	Malvaceae	Tropical Africa	Tropical Africa and South East Asia	Seed
Waltheria indica L.	Sterculiaceae	Tropical America	Tropical region of world	Seed
Xanthium strumarium	Asteraceae	Tropical America	Africa and Temperate and South East Asia	Seed
Youngia japonica (L.) DC.	Asteraceae	Tropical Asia	Worldwide	Seed

Table 1.
Invasive alien weed species world wide.

and spread to subtropical region of Asia, South America, Africa, Australia and Pacific Islands [25]. *Rubus fruticosus* L. (Family: Rosaceae) which is invasive weed is expected to retreat to subtropical and temperature regions and to higher altitude because sensitive to higher temperature and drought conditions [26]. *Nassella neesiana* (Trin and Rupr) Barkworth (Family: Poaceae) spread to new regions as it is highly invasive and drought resistant [27]. *Ulex europaens* L. (Family: Fabaceae) spread to high rainfall areas and cooler regions since the weed is drought sensitive. It is a weed in fifteen countries of the world from temperate to tropical areas and from coastal areas to mountains along a wide latitudinal and altitudinal gradient [28]. *Prosophis glandulosa* Torr. which belong to family Mimosaceae invade to warmer dry parts/lower rainfall areas because the weed is drought tolerant [29]. *Nassella trichtonia* (Nees) Hack. ex Arechav. weed belongs to family Poaceae spread to subtropical and temperate region and to higher altitude due to sensitive to temperature. It has diminished the agricultural carrying capacity of crops in south-eastern Australia, New Zealand and South Africa, and emerging populations have now been identified in Europe and the United States [30]. The changes in the distributions of globally noxious alien species (*Aegratina adenophora*, *Ageratum conyzoides*, *C. odorata*, *L. camara*, *M. micrantha*, and *P. hysterophorus*) in Bhutan, to provide evidence that even a mountain environment is under the threat of invasion given the change in climatic conditions which is a native of Central and South America [31]. *Ageratina adenophora* (Sprengel) R. King and H. Robinson (Asteraceae), is one of the most noxious invasive weeds in many parts of Asia, Oceania, and Africa. It has had serious ecological impacts on native biodiversity and caused enormous economic [32]. *Tagetes minuta* is a fast-growing annual weed that grows in moist and dry areas, from sea level to reasonable altitudes in the tropics and subtropics, and in soil pH ranging from 4.3 to 6.6. *Echium plantagineum*, an annual weed of the family Boraginaceae, is native to the western Mediterranean regions of Portugal, Spain and northern Africa, but is an introduced weed in the arid and temperate zones of Australia. *E. plantagineum* weed is a prolific seeder, producing up to 10,000 seeds per plant [33]. *P. hysterophorus* is a noxious weed in America, Asia, Africa and Australia and has now become one of the world's seven most devastating and hazardous weeds. P*arthenium*. *hysterophorus* alien weed is believed to have been introduced into India as contaminants in PL 480 wheat. P*arthenium. hysterophorus*, one of the most troublesome weeds in India and has also significantly expanded to Nepal. In Africa, there are about 35 invasive alien species were identified. Foremost among these are *P. hysterophorus* L., *E. crassipes* (Mart.) Solms, *Prosopis juliflora* (Sw.) DC., *L. camara* L., *Argemone ochroleuca* Sweet, *Xanthium strumarium* L., *A. conyzoides* L., *Datura stramonium* L., *Nicotiana glauca* Graham, *Senna didymobotrya* (Fresen.) Irwin & Barneby and *Senna occidentalis* (L.) which has spread from tropical and subtropical regions of South America. *A. ochroleuca* is flowering plants in the family Papaveraceae commonly known as prickly poppies and native to the West Indies and Central America; now a cosmopolitan tropical and subtropical weed. *S. didymobotrya* is a species of flowering plant in the Fabaceae (Leguminosae) which is native to Africa and found across the continents in several types of habitat [34]. *M. micrantha* has the largest distribution area (increase by 61–120%), while *adenophora* expand by 7–33%, *A. philoxeroides* by 12–74%, and *Ambrosia artemisiifolia* by 8–27%, respectively across globe. *A. adenophora*, Alternanthera philoxeroides, *A. artemisiifolia* and *M. micrantha* were invasive alien species to South East Asia native of Brazil. Invasive weed species, *L. camara*, A. adenophora, *P. hysterophorus* and *A. conyzoides* have reached 2900 m, which is higher than its reported elevation range (300–2800 m) across globe [35]. The distribution of invasive weed plants *A. adenophora*, *A. philoxeroides*, *A. artemisiifolia* and *M. micrantha* spreads towards the northern/southern ranges and higher elevation region worldwide due to susceptible to high temperature. Invasive weed species in family Poaceae (27 species), Asteraceae (23 species), Brassicaceae (18 species), Laminaceae (15 species), Fabaceae (11 species) and Caryophyllaceae (9 species) were recorded in the upper reaches of India [36].

3.3 Effect on crops by invasive alien weed species

Wide adaptability and faster growth of invaded weeds lead to dominance of weed in crop habitat Invasive weeds are responsible for 34% of agricultural losses [37] with the magnitude of impact varying between countries or location as 10% yield loss has been attributed to weeds in less developed countries and 25% in the least developed countries [38]. *Rottboellia cochinchinensis* is rated among the worst weeds in the world and is considered a serious problem in soybean (*Glycine max* (L.) Merr.), maize, cotton (Gossypium hirsutum L.), groundnut (*Arachis hypogaea* L.) and upland rice (*Oryza saliva* L.) in tropical regions of the world [39]. In tropical region it is a major weed problem in sugarcane (*Saccharum* spp.) and soybean. Invasive weed *Asphodelus fistulosus* a native of North America, South Europe and West Asia has been found in onion crop. The weed could make the land infertile if it is not controlled in a timely manner. *Imperata cylindrical* (L.) P. Beauv is one of the top invasive worst weed in the world and causes severe damage to the date palm and sugarcane fields of Iran [40].

Rice crop is infested with different invaded weed flora consisting of aquatic, semi-aquatic and terrestrial weeds (**Figure 3**). The invaded weed species *Alternanthera philoxeroides*, *Cyperus rotundus*, *Echinocloa crusgalli*, *Echinochloa stagnina*, *Eicchornia crassipes*, *Eragrostis stagnina*, *Commelina diffusa*, *Ludwigia liniflolia*, *Ageratum houstonianum*, *Alternanthera phiexeroides*, *Borrera articularis*, *Cynodon dactylon*, *Aeschynomene indica*, *Polygonum glabrum* Willd, *Melochia corchorifolia*, *Paspalam scrobiculatum*, and *Eleocharis acutangula* causes yield losses to the tune of 28–89 percent in transplanted and direct seeded lowland rice and 48–100 percent in upland

Echinochloa colona

Panicum repens

Eleusine indica

Echinochloa crusgalli

Paspalum distichum

Leersia oryzoides

Digitaria sanguinalis

Dactyactenium aeayptium

Dinebra retroflexa

Leptochloa uninervia

Cyperus rotundus

Cyperus iria

Fimbristylis dichotoma

Fimbristylus. meliacea

Scirpus microcarpus

Alternanthera echinata

Figure 3.
Invasive weed species in paddy lands.

ecosystem [41]. *Solidago gigantia* Aiton. had pronounced alleloapathic effect on germination and initial growth of carrot, barley and coriander. Reduction in emergence percent, shoot length and fresh weight of carrot and barley was also observed [42]. *Ageratum conyzoides*, *A. houstonianum* and *Erigeron karvinskianus* are primarily invading agroecosystem. *Avena fatua*, *Phalaris minor* and *Lolium temulentum* are the grassy weeds, which have now become a threat in wheat crop and affected yield [43].

Invaded weed species *P. minor* Retz, *Chenopodium album* L., *A. fatua* L., *Cichorium in*tybus, *Celosia argentia* and *Medicago denticulata* affect the yields in wheat [44]. *Echinochloa co*lona, *Trianthema portulacastrum*, *Euphorbia geniculata*, *Commelina communis* and *Physalis minima* invaded weed species affected soybean crop [45]. *Convolvulus arvensis* L., *Chicorium intybus* and *Lathyrus aphaca* invaded weed species affected chickpea crop. *Cynotis axillaris*, *Melochia conchorifolia* L., *Blainvillea acmella* (L.) Philipson (Asteraceae) and *Cyperus iria* native of Tropical America affected maize crop across globe [46]. *Tagetes minuta* is widely distributed across the tropics and subtropics and computing light, nutrients, and water with many economically important crops such as maize, rice, and beans. *Parthenium hysterophorus*, *Lantana camara*, *A. adenophora* and *A. conyzoides* are widely distributed and more rapidly proliferating alien plant weed species after crop yield [47]. Eighteen invasive weed species namely, *A. conyzoides*, *Cassia alata*, *Catharanthus pusillus*, *Celosia argentea*, *C. album*, *Eichhornia crassipes*, *Impatiens balsamina*, *Ipomoea eriocarpa*, *Ipomoea quamoclit*, *L. camara*, *Leucaena latisiliqua*, *Leucaena leucocephala*, *Melilotus alba*, *Mirabilis* *jalapa*, *Passifora foetida*, *Pennisetum purpureum*, *Portulaca oleracea* and *Prosopis julifora* have been introduced from South America affecting crops like rice, wheat, sorghum, oilseed and pulse crops in India [48]. *Echinochloa crus-galli*, *Setaria viridis* and *Digitaria sanguinalis* populations were high and *Sorghum halepens*, *Bidens* pilosa, *Acalypha wilkesiana*, *Galinsoga parviflora*, *Amaranthus retroflexus*, *Solanum physalifolium*, *C. album*, *Polygonum lapathifolium*, *Xanthium italicum Datura stramonium* and *Sicyos angulatus* affected maize, wheat, sunflower, sorghum, sugarbeet and soyabean crops [49]. Invasive weed species *Trifolium repens*, *Eryngium billardieri*, *Lemna minor* and *Sorgum halepense* are the major invasive weeds in hilly tracts of India affecting the yield of paddy, mustard, wheat and oats crops [50]. *Typha augustata* which belongs to the family Typhaceae is found across wetland ecosystem throughout the world affecting the yield of rice crop [51].

3.4 Measures to control invasive weed

Understanding invasive weed species ecology, morphology, reproductive biology, physiology, and biochemistry is essential for effective management and prevention management and control through a full range of factors regulating their density, growth and competitive ability. The weed management strategies could be adapted to minimize prevalence of the invasive species for reducing to minimize the undesired effects and optimizing land use by combining prevention and control practices [52]. Invasion by alien species in agroecosystem can be best controlled by measures like crop rotation, balanced fertilization, maintenance of cover crops, intercropping diversification, and alteration in soil physical chemical and biological properties.

Enforcement of strong legislation could prevent introduction of invasive alien weed species in the country for conserving the rich biodiversity and increase crop production. Prevention, early detection and eradication of invasive alien weed species is the most economical and effective means of management. It is important to ensure new weed species of vegetative reproductive weed parts are not introduced in new areas. Mechanical, physical, biological, and chemical (herbicide) have to be used for the control of invasive weed species across the world. Mechanical control usually refers to the mowing or mechanical cutting of an invasive plant infestation to limit seed production. Manual invasive plant control usually refers to hand-pulling or digging. Cultural control and competition including re-vegetating, irrigating or fertilizing to encourage the establishment of a healthy ground or crop cover to resist invasive plants. Biological control involves using living organisms to reduce seed production and vigor of an invasive plant species. Biological control agents are not available for many invasive plant species [53].

4. Conclusion

The twenty first century threat of invasive alien weed species is extensive and distributed globally. An invasion by alien weed species is a global problem and forms one of the major drivers of global change. Invasive weeds species are one of the major problems in crop production. The threat by invasive alien plant species has been with rapid growth of globalization. The species affect crop production and biodiversity. Apart from threat to biodiversity and ecological distribution invasive alien species have significant socio-economic impact. The weeds compete with crop plants for light, moisture, nutrients and space. The mechanism of plant weed invasions has been

DOI: http://dx.doi.org/10.5772/intechopen.106742

change in climatic condition, disturbance in natural ecosystem (soil, canopy cover, habitat fragmentation, fast growing potential of alien species and chemical interference by litter of alien weed species). The high seed production capacity spread, adaptation to wide climatic and soil condition are challenges to the management across worldwide for sustainable agricultural production.

Conflict of interest

The authors declare no conflict of interest.

Author details

Melekote Nagabhushan Arun[1*], Rapolu Mahender Kumar[1], Banugu Sreedevi[1], Guntupalli Padmavathi[2], Pallakonda Revathi[3], Neha Pathak[4], Dayyala Srinivas[1] and Boya Venkatanna[1]

1 Department of Agronomy, Indian Council of Agricultural Research - Indian Institute of Rice Research, Telangana, Hyderabad, India

2 Department of Plant Breeding, ICAR - Indian Institute of Rice Research, Hyderabad, Telangana, India

3 Department of Agronomy, PJTSU, Hyderabad, Telangana, India

4 Department of Botany, Aligarh Muslim University, Aligarh, India

*Address all correspondence to: arun_tulasi2011@yahoo.in

References

[1] Fleming JP, Wersal RM, Madsen JD, Dibble ED. Weak non-linear influences of biotic and abiotic factors on invasive macrophyte occurrence. Aquatic Invasions. 2021;**16**(2):349-364

[2] IPCC (Intergovernmental Panel on Climate Change). Climate Change 2021: The Physical Science Basis. Mitigation of Climate Change Synthesis Report. Cambridge, United Kingdom and New York, NY, USA: Cambridge University Press; 2021. p. 15. DOI: 10.1017/9781009157896

[3] Namkeleja SH, Thadeo T, Ndakidemi P. Allelopathic effects of Argemone mexicana to growth of native plant species. American Journal of Plant Sciences. 2014;**5**(9):1336-1344

[4] Hu CC, Lei YB, Tan YH, Sun XC, Xu H, Liu CQ, et al. Plant nitrogen and phosphorus utilization under invasive pressure in a montane ecosystem of tropical China. Journal of Ecology. 2019;**107**:372-386

[5] Uddin MDN, Robinson RW. Responses of plant species diversity and soil physical-chemical microbial properties to Phragmites australis invasion along a density gradient. Scientific Reports. 2017;**7**:11007

[6] Sardans J, Bartrons M, Margalef O, Gargallo-Garriga A, Janssens IA, Ciais P, et al. Plant invasion is associated with higher plant-soil nutrient concentrations in nutrient-poor environments. Global Change Biology. 2017;**23**:1282-1291

[7] Funk JL. The physiology of invasive plants in low-resource environments. Conservation Physiology. 2013;**1**:1-17. DOI: 10.1093/conphys/cot026

[8] Sun S, Chen J, Feng W, Zhang C, Huang K, Guan M, et al. Plant strategies for nitrogen acquisition and their effects on exotic plant invasions. Biodiversity Science. 2021;**29**:72-80

[9] Shannon-Firestone S, Reynolds HL, Phillips RP, Flory SL, Yannarell A. The role of ammonium oxidizing communities in mediating effects of an invasive plant on soil nitrification. Soil Biology and Biochemistry. 2015;**90**:266-274

[10] Rossiter-Rachor NA, Setterfield SA, Douglas MM, Hutley LB, Cook GD, Schmidt S. Invasive Andropogon gayanus (Gamba grass) is an ecosystem transformer of nitrogen relations in Australian savanna. Ecological Applications. 2009;**19**:1546-1560

[11] Li T, Li Q, Xiong P, Luo Y, Zhang X, Lin GB. Direct and indirect effects of environmental factors, spatial constraints, and functional traits on shaping the plant diversity of montane forests. Ecological and Evolution. 2020;**10**(1):557-568

[12] Broadbent A, Stevens CJ, Peltzer DA, Ostle NJ, Orwin KH. Below ground competition drives invasive plant impact on native species regardless of nitrogen availability. Oecologia. 2018;**186**:577-587

[13] Bhowmik PC. Invasive weeds and climate change: Past, present and future. Journal of crop and weed. 2014;**10**(2):345-349

[14] Early R, Bethany AB, SDukes J, Joshua JL, Julian DO, Dana MB, et al. Global threats from invasive alien species in the twenty-first century and national response capacities. Nature

Communications. 2016;**7**:12485. DOI: 10.1038/ncomms12485

[15] Amare T. Review on impact of climate change on weed and their management. American Journal of Biological and Environmental Statistics. 2016;**2**(3):21-27. DOI: 10.11648/j.ajbes.20160203.12

[16] Ziska LH. The role of climate change and increasing atmospheric carbon dioxide on weed management: Herbicide efficacy. Agriculture Ecosystems and Environment. 2016;**231**:304-309

[17] Bajwa AA, Chauhan BS, Adkins SW. Germination ecology of two Australian biotypes of ragweed parthenium (Parthenium hysterophorus) relates to their invasiveness. Weed Science. 2018;**66**:62-70

[18] Clements DR, Jones VL. Rapid evolution of invasive weeds under climate change: Present evidence and future research needs. Frontiers in Agronomy. 2021;**3**:664034. DOI: 10.3389/fagro.2021.664034

[19] Khanduri A, Biswas S, Vasistha HB, Rathod D, Jha SK. A status of invasive alien species plant diversity in Tehri district forest ecosystem of Garhwal Himalayan region. Current World Environment. 2017;**12**(2):377-388

[20] Ziska LH, McClung A. Differential response of cultivated and weedy (red) rice to recent and projected increases in atmospheric carbon dioxide. Agronomy Journal. 2008;**100**:1259-1263

[21] Ehrenfeld JG. Ecosystem consequences of biological invasions. Annual Review of Ecology, Evolution and Systematics. 2010;**41**:59-80

[22] Ordonez A, Wright IJ, Olff H. Functional differences between native and alien species: A global-scale comparison. Functional Ecology. 2010;**24**:1353-1361

[23] Rusdy M. Imperata cylindrica: Reproduction, dispersal, and controls. CAB Reviews. 2020;**15**:1-9

[24] Sekar KC. Invasive alien plants of Indian Himalayan region—Diversity and implication. American Journal of Plant Sciences. 2020;**3**(2):177-184. DOI: 10.4236/ajps.2012.32021

[25] Smith LM, Reynolds HL. Light, allelopathy, and post-mortem invasive impact of garlic mustard on native forest understory species. Biological Invasions. 2014;**16**:1131-1144

[26] Clements DR. Invasive weed species and their effects. In: Zimdahl R, editor. Integrated Weed Management for Sustainable Agriculture. Cambridge, UK: Burleigh Dodds Science Publishing; 2017. pp. 65-88

[27] Panetta FD, Gooden B. Managing for biodiversity: Impact and action thresholds for invasive plants in natural ecosystems. NeoBiota. 2017;**34**:53-66. DOI: 10.3897/neobiota.34.22821

[28] Champika SK, Sujith SR. Reproductive biology of Ulex europaeus Fabaceae in the mount lofty ranges of South Australia and Srilanka. The International Journal of Plant Reproductive Biology. 2019;**11**(2):145-152

[29] Ahmed MA, Sameh KAE, Soad MEA, Wagdi SS, Noa ET. Jesus MC capability of the invasive tree Prosopis glandulosa Torr. To remediate soil treatment with sewage sludge. Sustainability. 2019;**11**:1-13. DOI: 10.3390/su11092711

[30] Humphres T, Dowling K, Turville C, Sinclair S, Florentive S. Ecology, distribution and control of

the invasive weed Nassella trichotoma (Nees) hack ex Arachav. A global review of current and future challenges. Weed Research. 2020;**60**(6):392-405

[31] Thiney U, Banterng P, Gonkhamdee S, Katawatin R. Distributions of alien invasive weeds under climate change scenarios in mountainous Bhutan. Agronomy. 2019;**9**(8):1-16. DOI: 10.3390/agronomy9080442

[32] Poudel AS, Jha PK, Shrestha BB, Muniappan R. Biology and management of the invasive weed Ageratina adenophora (Asteraceae): Current state of knowledge and future research needs. Weed Research. 2019;**59**:79-92

[33] Florentine S, Weller S, King A, Florentine A, Dowling K, Westbrooke M, et al. Seed germination response of a noxious agricultural weed Echium plantagineum to temperature, light, pH, drought stress, salinity, heat and smoke. Crop and Pasture Science. 2018;**69**:326-333

[34] Tamiru G. Invasive alien weed species distribution, impacts on agriculture, challenge and reaction in Ethiopia: A review. Journal of Biology, Agriculture and Healthcare. 2017;7(7):136-146

[35] Rai PK, Singh MM. Lantana camara invasion in urban forests of an indo-Burma hotspot region and its eco-sustainable management implication through bio monitoring of particulate matter. Journal of Asia-Pacific Biodiversity. 2015;**8**(4):375-381

[36] Dar PA, Reshi ZA. Assessment of plant invasions in agroecosystems of Kashmir Himalaya for better management. Frontiers in Agronomy. 2022;**3**:788-797. DOI: 10.3389/fagro.2021.788797

[37] Inderjit Pergl J, van Kleunen M, Hejda M, Babu CR, Majumdar S, Singh P, et al. Naturalized alien flora of the Indian states: Biogeographic patterns, taxonomic structure and drivers of species richness. Biological Invasions. 2018;**20**(6):1625-1638

[38] Radicetti E, Mancinelli R. Sustainable weed control in the agro-ecosystems. Sustainability. 2021;**13**:8639. DOI: 10.3390/su13158639

[39] Paini DR, Sheppard AW, Cook DC, De Barro PJ, Worner SP, Thomas MB. Global threat to agriculture from invasive species. Proceeding National Academy of Sciences of the Unites States of America. 2016;**113**:7575-7579

[40] Funez LA, Ferreira JPR, Hassemer G, Trevisan R. First record of the invasive species Rottboellia cochinchinensis (Poaceae, Andropogoneae) in the south region of Brazil. The Journal of Biodiversity Data. 2016;**12**(4):1-4. DOI: 10.15560/12.4.1930

[41] Haidar H, Seyed E, Al-Alahmadi M. Effect of environmental factors on rhizome bud germination and shoot emergence of invasive Imperata cylindrica. Weed Research. 2021;**61**(5):374-384

[42] Rai PK. Concept of plant invasion ecology as prime factor for biodiversity crisis: Introductory review. International Research Journal of Environment Sciences. 2015;**4**(5):85-90

[43] Balicevic R, Ravlic M. Tea Zivkovic Allelopathic effect of invasive species giant goldenrod (Solidago gigantia Ait.) on crops and weeds. Herbologia. 2015;**15**(1):19-29

[44] Amare T. Review on impact of climate change on weed and their management. American journal of

biological and environmental. Statistics. 2016;**2**(3):21-27

[45] Shahzad M, Farooq M, Hussain M. Weed spectrum in different wheat-based cropping systems under conservation and conventional tillage practices in Punjab, Pakistan. Soil and Tillage Research. 2016;**163**:71-79

[46] Raj R, Das TK, Kaur R, Singh R, Shekhawat K. Invasive noxious weed management research in India with special references to Cyperus rotundus, Eicchornia crassipes and Lantana camara. Indian Journal of Agricultural Sciences. 2018;**88**(2):181-196

[47] Matsuhashi S, Asai M, Fukasawa K. Estimations and projections of Avena fatua dynamics under multiple management scenarios in crop fields using simplified longitudinal monitoring. PLoS One. 2021;**16**(1):26-31. DOI: 10.1371/journal.pone.0245217

[48] Matshwene EM, Solomon WN. Mexican poppy (Argemone mexicana) control in corn field using deep learning neural networks: A perspective, Acta Agriculturae Scandinavica. Section B — Soil and plant. Science. 2019;**69**(3):228-234. DOI: 10.1080/09064710.2018.1536225

[49] Pathak R, Vikram SN, Ranbeer SR, Indra DB. Alien plant invasion in the Indian himalayan region: State of knowledge and research priorities. Biodiversity and Conservation. 2019;**28**:3073-3102

[50] Peters K, Breitsameter L, Gerowitt B. Impact of climate change on weeds in agriculture: A review. Agronomy for Sustainable Development. 2014;**34**:707-721

[51] Tabe Ojong MP, Alvarez M, Ihli HJ. Action on invasive species: Control strategies of Parthenium hysterophorus L. on small holder farms in Kenya. Environmental Management. 2022;**69**:861-870. DOI: 10.1007/s00267-021-01577-5

[52] Bansal S, Lishawa SC, Newman S, et al. Typha (cattail) invasion in north American wetlands: Biology, regional problems, impacts, ecosystem services, and management. Wetlands. 2019;**39**:645-684. DOI: 10.1007/s13157-019-01174-7

[53] Li W, Luo J, Tian X, et al. A new strategy for controlling invasive weeds: Selecting valuable native plants to defeat them. Scientific Reports. 2015;**5**:11004. DOI: 10.1038/srep1100

Chapter 6

Perspective Chapter: Plant Invasion and Ecosystem Litter Decomposition

Nasir Shad, Zohra Nasheen, Rabia Afza and Ling Zhang

Abstract

Litter decomposition plays an important role in the biogeochemical cycling of elements in ecosystems. Plant trait differences especially between invasive and native species lead to changes in litter decomposition rates. The litter decomposition rate is influenced by climatic factors such as seasonal variations, humidity, temperature, and rainfall, where species litter may have different responses. This review aims to better understand how litter decomposes in ecosystems associated with plant invasion and global changes. It also reviews the effects of various factors on litter degradation as well as how quickly invasive litter decomposes and contributes to greenhouse gases (GHGs) emissions. Single species litter or only aboveground litter studies may not sufficiently represent ecosystem dynamics; therefore, the co-determination of above- and belowground litter in a mixture of species diversity is required in different biomes interaction with global change factors. As a result, comprehensive litter degradation studies must be conducted in order to understand the turnover rate of nutrients and other elements in these sensitive ecosystems.

Keywords: litter traits, invasion, ecosystem, decomposition, GHG

1. Introduction

Litter decomposition is an important process of nutrient cycling in ecosystems. It releases various elements into the soil back to support the plant. The root system absorbs nutrients and provides a source, realizing the exchange of chemical elements within the ecosystem. It is estimated that more than 90% of the nutrients absorbed by plants of nitrogen and phosphorus and more than 60% of mineral elements are returned to the soil through litter degradation hence biogeochemical cycling [1]. Moreover, many studies focused on aboveground litter and few studies found on belowground litter decomposition; however, the combination of both and their environmental condition may well define the ecosystem litter decomposition dynamics.

Invasive species-associated traits have impacts on litter input and litter quality, which possibly alter soil chemistry and change in microbial biomass and activities influences by climatic changes, monitoring the litter decomposition rate dynamics [2–5]. Litter traits of invasive species are leading factor determining the influences of decomposition rate [2, 6]. Invasive species may get advantage over native species due to higher-quality litter from invasive species. In addition, invasive species have fast growth, higher leaf traits, leaf toughness, and high-nutrient content governing decomposition rate. However, decomposition litter's non-additive effects may differ from single species decomposition.

Plant invasion and their associated traits-enhanced decomposition rate are also predicted to alter GHGs emission to the atmosphere [7, 8]. Plant invasion alter GHGs emission through its litter input, quality, and changes in soil microbes [9, 10]. However, global change factors through interaction WITH plant invasions indirectly or directly affect litter decomposition. Elevated carbon dioxide (CO_2), nitrogen (N) deposition, and ultraviolet B (VU-B) radiation have been reported to alter decomposition rate of invasive plants [7, 11]. The data presented in this chapter were obtained from professional websites and diverse databases. We conducted a survey in the Web of Science, Google Scholar, PubMed, Science Direct, Springer, CAB abstracts, Taylor, and Francis using different keywords including plant invasion, litter decomposition, GHGs, decomposer, etc., to obtain relevant information regarding litter decomposition in the context of plant invasion and global climate change.

2. Litter decomposition in natural ecosystems

Ecologically, litter refers to dead plant materials or detached from a living plant [9]. Aboveground plant organs (i.e., stem, leaves, and reproductive organs) and belowground plant organs (i.e., roots) form plant litters. These litters decompose through biological (soil microorganism such as bacteria and fungi), physical (abiotic forces such as wind, temperature, moisture, light), and chemical processes, converting organic matter into nutrients and CO_2 [12], hence, maintaining soil nutrients and soil fertility, by vigorously giving support to plant diversity [13]. In turn, microorganism heterotrophic respiration returns CO_2 to the atmosphere [14]. Flora is significantly influenced by litter decomposition, where litter adds nutrients to the ecosystem as recycling of nutrients. Litter decomposition rates are dynamics affected by microbes and soil fauna influences by climatic conditions, rainfall, temperature, and seasonal fluctuations. Litter chemistry is also involved, which influences the decomposition process and soil microbial activities [15]. Slow decomposition adds nutrients stocks and organic matter to the soil; however, fast decomposition rates offer more nutrients to meet plant intake requirements [16]. Additionally, litter chemistry (i.e., organic compounds, N content, C/N ratio), which varies to plant organs (leaves, stem, root, etc.) and plant species (i.e. native vs. invasive), influences the litter quality and decomposition rate [9, 17]. Litter decomposition varies at species level (i.e., native vs. invasive), predicting that species with high N and ash content as well as low C/N and lignin content resulted in high litter decomposition rate [18]. Above- and belowground litters have different environmental conditions where the temperature and precipitation play a key role in difference between leaf and root litter decomposition. Decomposer organisms are another factor, and their biomass decreases in shift community with increase in soil depth. The coordination between leaf and fine root litter decomposition may weaken by divergent decomposition position, yet their mechanism and differences in regulating factors are still unexplored.

2.1 Aboveground litter

Aboveground litter is an organic horizon originated by the plant materials on soil surface [19]. This aboveground organic horizon formation depends on the litterfall rate and decomposition rate of plant materials [20]. Leaves are the main component of forest litter, accounting for about 70%, contributing to more than any aboveground plant organs. In addition, leaves contributed more to nutrient budget in short term, and its decomposition is faster than any other plant organ in grassland and forest [21]. The functional trait syndromes of coarse stem components may not be coordinated with those of other organs [22]; as a result, their afterlife effects on decomposability are poorly coordinated [23]. However, its decomposition rate and turnover may be slow, which is less important in short-term nutrient cycling. Nevertheless, forests have a lot of coarse woody debris (such as branches, stumps, and coarse roots), and leaf decomposition rates alone are not enough to forecast organic matter dynamics [24, 25]. The aboveground litterfall may be influenced by seasonal behavior, characteristics of plantation/species such as aboveground biomass, volume, canopy closure [26], and the environmental condition of the region [27]. On the other hand, litter chemistry and climatic condition play a key role in decomposition rate [28]. Such regulating factors define the differences in aboveground versus belowground litter decomposition, being a gap in the development of ecological research.

2.2 Belowground litter

Most studies have shown that the decomposition rate of root litter is significantly lower than that of leaf litter [21, 29, 30]. The decomposition environment of root litter is very different from that of aboveground litter, and the regulating factors are different [31–33]. Therefore, the decomposition rate of leaves and its controlling factors cannot be used to infer root decomposition. Sun et al. [34] 6-year findings on 35 species show that based on the rate of decomposition, leaves litter (77%) was higher than root litter (35%), with different regulating factors where leaves decomposition is controlled mainly by lignin:nitrogen ratio; however, non-lignin carbon compounds (phenols and tannins) played a dominant role in root decomposition. In addition, there is no correlation between fine root and leaf litter decomposition rates among different tree species. Moreover, the root size in diameter that may also play an important role in decomposition rate is still controversial [35]. For a long time, due to the limitations of technology and methods, the research of underground ecosystem has become a bottle that restricts the development of ecology [36]. The decay and decomposition of the root system are of great significance to the carbon cycle and the availability of nutrients in the soil. Therefore, in the future, research on the decomposition of underground root litter (such as aboveground and underground whole, roots of different diameters) should be strengthened, especially influencing control mechanism under the background of global change (**Figure 1**).

3. Factors impacting litter decomposition

Litter decomposition occurs by three major process: fragmentation, leaching, and catabolism [39–41]. Leaching is the process of removing soluble materials from degrading organic matter. Fragmentation occurs by soil fauna or abiotic agent produces substrate that benefits soil microbes. The catabolic activity of bacteria and fungi is mostly responsible for the chemical change of dead organic matter. Hence, the litter decomposition is regulated by three main factors: litter quality (litter physiochemical characteristics), physicochemical environment (abiotic), and decomposer organism (biotic) [42]. These factors are described in **Figure 2**.

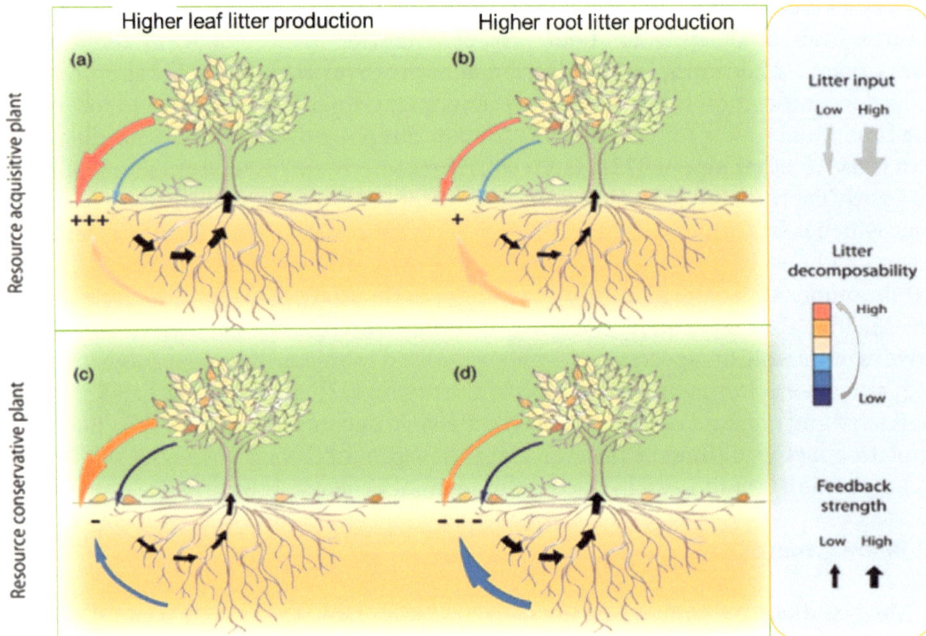

Figure 1.
Litter inputs (above- and belowground) co-determine the quality, quantity of litter, and decomposability; and therefore, the dynamics of biogeochemical cycle. A resource acquisitive plant community (a, b) produces litter of consistently higher decomposability than a resource conservative plant community (c, d) (modified from Freschet et al. [21]). In addition, the majority of invasive species in position of higher resource acquisition than native species (see collection of studies [37, 38]), but this may vary or not be the same in regions. Therefore, invasive species may fit to the resource acquisitive plant model to determine their composability but regional differences and environmental factors may influence litter decomposition rate.

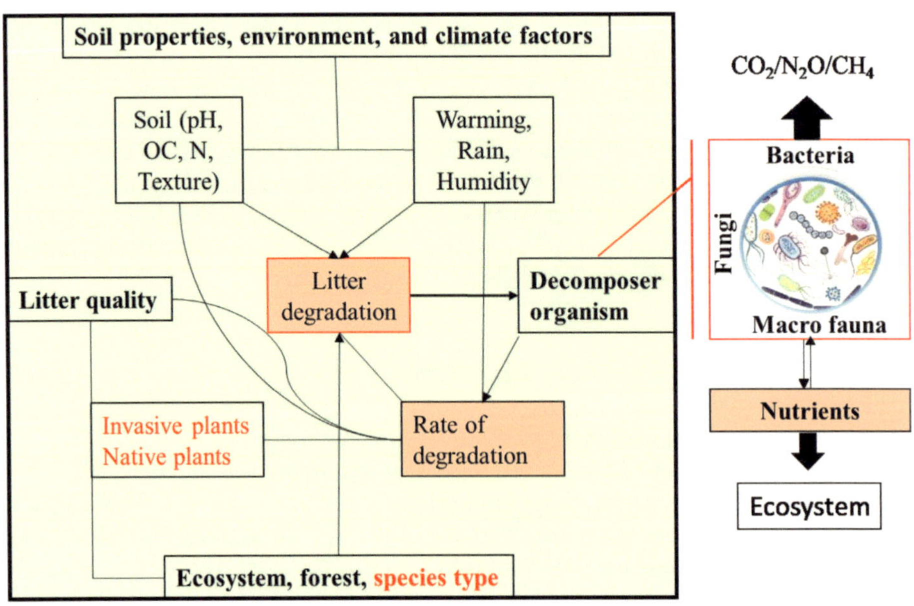

Figure 2.
Various factors affecting litter degradation (modified from Krishna and Mohan [9]).

3.1 Litter quality

Litter quality can be explained by litter chemistry, which determines litter decomposition [43], and altering soil biota facilitates the decomposition of litter materials [44]. Hence, litter quality such as chemical properties, N concentration, as well C/N ratio controlled the release of nutrients and litter decay [45–47]. Studies investigating litter quality impacting decomposition are common [48], where C:N ratio and N content play a leading role followed by lignin:N ratio [48–51]. Nevertheless, the existence of other regulating factors (soil biota, environment) interacting with litter quality makes complexity and is poorly understood. These variables and their impact on litter decomposition depend on plant species and soil characteristics. The quantity and quality of litter constituents, the chemical and physical environment, and decay entities are the key factors that influence organic matter conversion. Moreover, decomposer arrangement in the soil influences litter decomposition rate and nutrient dynamics [42].

3.2 Decomposer

The rate of litter decomposition is known to be affected by the abundance and arrangement of soil fauna and microbial communities at distinct stages of decomposition [52]. Organic material decomposition has significant role in nutrient cycling and energy flow in the ecosystem. Earlier research has identified the importance of a variety of bacteria and fungi in litter decomposition [9], which demonstrated that forest soil and related microbial communities play an important role in the decomposition of litter, under laboratory conditions. Fungi are the top decomposers in the soil microfauna, with a 75% better ability to decompose organic materials than other microbes [53]. Besides, bacteria that involved in litter decomposition mineralization process account for 25–30% of the total microbial biomass in the soil [54]. Generally, decomposers usually grow once the litter reaches the ground; however, the growth of microbes on the litter, particularly fungi, may start decomposition before the litter falls. The composition of microbial community that lives in the litter is determined by the litter qualities, soil characteristics, and changes in these characteristics through time [55]. The quantity and quality of litter input dependent on plant species also affects litter decomposition [4].

Beside bacteria and fungi, soil fauna (micro- and macro-invertebrates) because of their important involvement in mineralization processes, organic matter decomposition and nutrient cycling, as well as pedogenesis, which persist in the litter strata and on the soil upper layer, they are an important aspect of ecosystems [56]. Soil faunal activities primarily aid in the acclimatization of litter and the stimulation of microbial activity.

3.3 Soil and climate factors

Litter decomposition is influenced by the physical and chemical features of the soil. Texture is the most important of them all because it affects water and nutrient dynamics, porosity, permeability, and surface area [9]. Organic matter content, pH, nutrients, and cation exchange capacity are some of the most important chemical features. Among them, organic matter influences the major soil properties, such as soil physio-chemistry (pH, bulk density) affecting litter decomposition, also increasing microbial density [57]. Soil N is considered the primary regulating factor and had received worldwide attention, while phosphorus act as limiting nutrients because of poor availability and circulation in major forests. Furthermore, the rate of litter decomposition is influenced by soil temperature and N concentration. Although magnesium and potassium are important minerals for higher plants, they have little effect on microbial activity and are quickly eliminated from decomposing litter.

Temperature, moisture, and other climatic conditions may influence the rate of litter decomposition. Several studies recorded slow and fast decomposition rate in winter and rainy season, respectively [58–60], and the higher decomposition in rainy season may be due to the higher micro-fungal population, which may increase with suitable moisture and sufficient rainfall. Zhang and Wang [28] examined 785 datasets demonstrated that mean annual temperature was important factor driving decomposition on global scale at different climatic zones. Kumar et al. [61] also found increase in litter decomposition rate and weight loss in rainy season with high microbial load and soil moisture. However, there is still controversy about which climate index best predicts decay rates. Although there is water in the soil, actual evapotranspiration is the most important factor in determining the litter decomposition rate [62]. In contrast, some studies disagreed with the concept of relation between litter decomposition and actual evapotranspiration as a reliable indicator of decay litters [63–65].

4. Plant invasion effects on litter input and decomposition rate

Increase in invasive species traits such as specific leaf area, leaf dry matter content, plant height and stem-specific density [66] accumulates high biomass, and nutrient contents, especially N, P, and K [67], produce high quantity, and quality litter further increases litter decomposition rate. Invasive plant litter input further increases microbial biomass [67], and soil enzymatic activities [67, 68] may be attributed to the fast decomposition rates.

Plant invasions in the invaded ecosystems can significantly change the ecosystem functioning. Invasive alien plants may have a significant impact on litter decomposition, which is one of these ecosystem services. According to reports, the C/N ratio can be used to calculate the decomposition of leaf litter, which varies at species level [69]. High-quality leaves (nutrient-rich leaves) decompose at a faster rate than low-quality leaves (nutrient-deficient leaves). Invasive species produce litter of higher chemical quality, which decompose rapidly and release high nutrients to the soil [70]. In general, species with high ash and nitrogen content, as well as low C/N ratios and lignin concentration, decompose quickly [71]. Patil et al. [70] found higher N and P concentration, and low lignin and C/N ratio in invasive species resulted in higher rate of litter decomposition when compared to native species. Several studies found that litter nitrogen concentration and the C/N ratio are closely linked to litter degradation rates [72]. Phosphorus concentrations and C/P ratio seemed to be strong indicators of degradation rate [73]. Plant litter lignin concentrations and lignin/N ratios are also good predictors of litter decomposition rate [69].

A meta-analysis of 94 studies across world by Liao et al. [74] demonstrated that invaded communities have a 117% rise in litter decomposition rates. In contrast, some findings revealed no change or even decrease in litter decomposition rate associated with invasive plants [3, 74, 75]. These effect differences of invasive species on litter decomposition rates may influence functional (litter) traits of invasive species and the communities [76]. Thus, functional traits can be a good and mechanistic approaches define variation in decomposition rates. Leaf traits have been found to affect litter quality and hence alter litter decomposition rates at species level [2, 77, 78]. Invasive species leaf traits are associated with fast growth, high specific leaf area, leaf thickness and toughness, low leaf dry matter content, and high leaf nutrient concentrations, and such traits are directly involved in high leaf decomposition rates [66, 79]. Study by Zhang et al. [80] found that litters of invasive *Alternanthera philoxeroides* decompose faster than annual native grass (*Eragrostis pilosa*), but in case of mixture, invasive litter decomposition rate was constant and caused increase in the native litter decomposition rate.

Single species do not sufficiently represent an ecosystem function and litter decomposition. Non-additive effects mainly depended on species composition and diversity may further strengthen by litter chemistry [81]. Mixed litter can change the chemical environment as well as the physical surface area of the litter where decomposition takes place. It can also have an impact on ecological processes including soil respiration [82], net N mineralization [83], and microbial activity in the soil. Invasive *Solidago canadensis* litter alone had higher N concentration, but in mixture with native *Phragmites australis* decreased the N concentration of *Solidago canadensis* litter but increased that of *Phragmites australis* litter [83]. Nutrient transfer across litters in mixture may have favorable non-additive effects [84]. The single species litter decomposition is possibly to shift by litter mixture, and that mixture litter effect has been observed as positive interaction in early stage and been shift to negative non-additive effect over certain time duration [83, 85]. Litter decomposition of invasive species is more depended on N and P than native species [6], and in a case of N transfer of invasive litter to native litter may slow down the decomposition rate, which may be influenced by secondary metabolites.

5. Plant invasion effects on soil GHGs emissions via litter decomposition

The raising GHGs disrupt the natural ecosystems and challenge the management of natural habitats on a global scale. The anthropogenic activities are directly or indirectly involved in the production of these GHGs emissions. Plant invasion is one of the factors that alter GHGs emissions in many ways mainly by their litter quantity, quality, and soil microbe alteration involved in the production of GHGs emissions/process [9]. The numerous elements and chemicals that accumulate in plants are released into the environment via litter decomposition, leaching into the soil and diffusing into the atmosphere. As a result, these compounds are being reintroduced into the biogeochemical cycle. Soil chemical environment created by litters associated with invasion may shift biogeochemical processes through changes in litter decomposition rate. Grassland ecosystems invaded by *Alternanthera philoxeroides or Solidago canadensis* produced higher N_2O (60%) and CO_2 (30%) than that of non-invaded grassland ecosystems (dominated by *Eragrostis pilosa* or *Sesbania cannabina*) [86], where invaded ecosystem, produced more (155–361%) rapidly decomposing litter than that of non-invaded ecosystem [82], might be due the large litter input and higher C:N ratio associated with invasions. Another study by Zhang et al. [83] demonstrated that *Solidago canadensis* invasion loss (mass and N) was higher than native *Phragmites australis*, while native decreased and invasive increased N loss in a mixture litter (**Figure 3**).

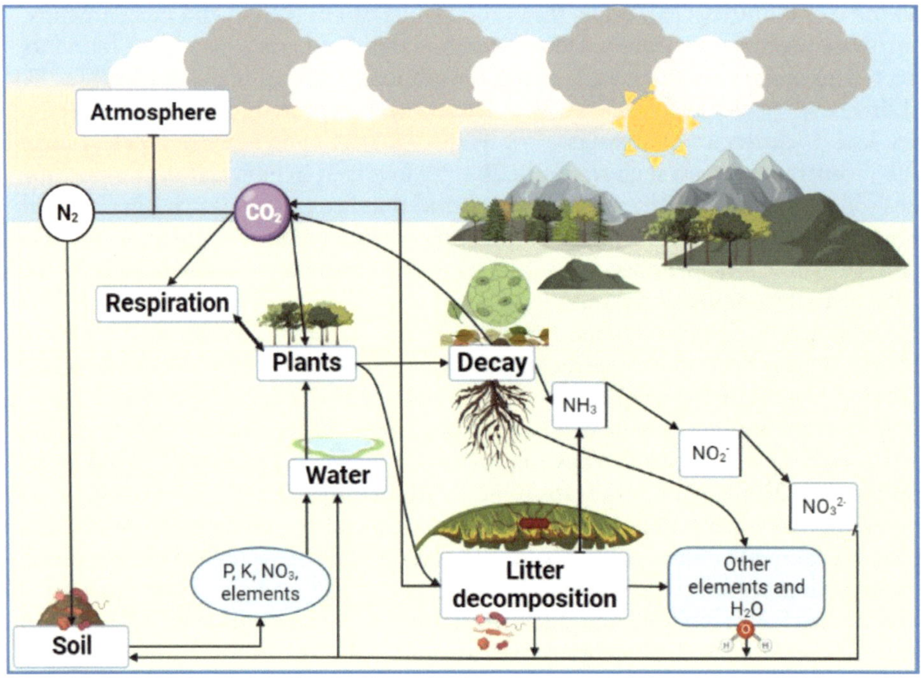

Figure 3.
Litter decomposition and its role in biogeochemical cycle. The accumulated compounds and elements by plants returns to environment through litter degradation; conversion of these organic matter results in nutrients and GHGs into the atmosphere (modified from Krishna and Mohan [9], created with biorender, https://www.biorender.com/).

6. Interactive effects between plant invasions and other global change factors on litter decomposition

Biological invasions-associated traits alter soil chemistry [87], soil enzymatic activities [68], further impact on litter decomposition [7], which resulted in higher GHGs emission [88]. Nitrogen deposition and elevated CO_2 potentially affect litter decomposition through change in litter quality and quantity [7]. In contrast, elevated CO_2 decreased the invasive *Triadica sebifera* litter decomposition via change in litter quality [7]. Moso bamboo invasion decreased the sensitivity of decomposition rate in response to mean annual precipitation and mean annual temperature [89].

An increase in water level caused decrease in nutrient release and decomposition rate from invasive plant litter [90], which might be due to the decrease of anaerobic microbial activities in high water level [91]. Suitable moisture condition will offer more nutrients that increase microbial biomass and activities will further accelerate litter decomposition rate [92]; invasive species often associated with nutrient-rich litter will get an advantage in this case.

UV-B exposure during litter decomposition may have a direct effect on decomposition rates by changing photodegradation states or decomposer composition in the litter, whereas UV-B exposure during growth periods may alter plant chemistry and physical properties. Meta-analysis of six biomes by Song et al. [93] demonstrated that elevated VU-B directly (7%) and indirectly (12%) increased litter decomposition rate, while attenuated UV-B decreased the rate of litter decomposition. In addition, the interactive effect of UV-B and N deposition increased litter decomposition rate of moso bamboo [94], and coarse woody debris [11].

7. Conclusions

This review focuses on ecosystem litter decomposition in the context of plant invasion associated with global change factors. Litter quality and traits in association with soil microbes and environmental factors play a leading role in decomposition rate. Plant traits associated with plant invasion alter litter input, soil microbes, and soil enzymatic activity, and hence increase litter decomposition rates. Studies on single species, and/or only aboveground litter or belowground litter do not represent decomposition rate sufficiently in ecosystem. Therefore, there is required co-determination of species diversity in different biomes both above- and belowground litter with their associated environment, regulating factors, and global change factors.

Acknowledgements

The study was fanatically supported by Research Funding of Jiangxi Agricultural University (9232305172) and First-Class Discipline of Forestry of Jiangxi Province.

Conflict of interest

The authors declare no conflict of interest.

Author details

Nasir Shad[1,3], Zohra Nasheen[2], Rabia Afza[3] and Ling Zhang[1*]

1 Key Laboratory of Silviculture, College of Forestry, Jiangxi Agricultural University, Nanchang, China

2 Key Laboratory of Vegetation Ecology, Northeast Normal University, Changchun, China

3 Department of Botany, Laboratory of Molecular Biology and Tissue Culture, Hazara University Mansehra, Pakistan

*Address all correspondence to: lingzhang09@126.com

References

[1] Chapin FS, Matson PA, Mooney HA, Vitousek PM. Principles of Terrestrial Ecosystem Ecology. New York: Springer; 2002

[2] Cornwell WK, Cornelissen JH, Amatangelo K, Dorrepaal E, Eviner VT, Godoy O, et al. Plant species traits are the predominant control on litter decomposition rates within biomes worldwide. Ecology Letters. 2008;**11**(10):1065-1071

[3] Ehrenfeld JG. Effects of exotic plant invasions on soil nutrient cycling processes. Ecosystems. 2003;**6**(6):503-523

[4] Hättenschwiler S, Tiunov AV, Scheu S. Biodiversity and litter decomposition in terrestrial ecosystems. Annual Review of Ecology, Evolution, and Systematics. 2005;**36**:191-218

[5] Fang H, Liu Y, Bai J, Li A, Deng W, Bai T, et al. Impact of Moso bamboo (*Phyllostachys edulis*) expansion into Japanese cedar plantations on soil fungal and bacterial community compositions. Forests. 2022;**13**(8):1190

[6] Godoy O, Castro-Díez P, Van Logtestijn RS, Cornelissen JH, Valladares F. Leaf litter traits of invasive species slow down decomposition compared to Spanish natives: A broad phylogenetic comparison. Oecologia. 2010;**162**(3):781-790

[7] Zhang L, Zou J, Siemann E. Interactive effects of elevated CO_2 and nitrogen deposition accelerate litter decomposition cycles of invasive tree (*Triadica sebifera*). Forest Ecology and Management. 2017;**385**:189-197

[8] Yao Y, Song Y, Su P, Wang J, Miao C, Luo Y, et al. Asymmetric responses of functional microbes in methane and nitrous oxide emissions to plant invasion: A meta-analysis. Soil Biology and Biochemistry. 2023;**178**:108931

[9] Krishna M, Mohan M. Litter decomposition in forest ecosystems: A review. Energy, Ecology and Environment. 2017;**2**(4):236-249

[10] Lai X, Luo L, Fang H, Zhang L, Shad N, Bai J, et al. Bacterial and fungal inhibitor interacted impacting growth of invasive *Triadica sebifera* and soil N_2O emissions. Frontiers in Forests and Global Change. 2023;**6**:121

[11] Wu C, Wang H, Mo Q, Zhang Z, Huang G, Kong F, et al. Effects of elevated UV-B radiation and N deposition on the decomposition of coarse woody debris. Science of the Total Environment. 2019;**663**:170-176

[12] Aerts R. Climate, leaf litter chemistry and leaf litter decomposition in terrestrial ecosystems: A triangular relationship. Oikos. 1997;**79**(3):439-449

[13] Ma L, Guo C, Xin X, Yuan S, Wang R. Effects of belowground litter addition, increased precipitation and clipping on soil carbon and nitrogen mineralization in a temperate steppe. Biogeosciences. 2013;**10**(11):7361-7372

[14] Schimel DS. Terrestrial ecosystems and the carbon cycle. Global Change Biology. 1995;**1**(1):77-91

[15] Chapman SK, Koch GW. What type of diversity yields synergy during mixed litter decomposition in a natural forest ecosystem? Plant and Soil. 2007;**299**(1):153-162

[16] Isaac SR, Nair MA. Biodegradation of leaf litter in the warm humid tropics of Kerala, India. Soil Biology and Biochemistry. 2005;**37**(9):1656-1664

[17] Rahman MM, Petit-Aldana J, Tsukamoto J, Wu Q-S. Litter chemistry, decomposition and its effects on soil biogeochemistry of forest ecosystems. In: Elliott D, editor. Forest Ecosystems: Management, Impact Assessment and Conservation. New York, USA: Nova Publication; 2017. pp. 21-44

[18] Hornbach DJ, Shea KL, Dosch JJ, Thomas CL, Gartner TB, Aguilera AG, et al. Decomposition of leaf litter from native and nonnative woody plants in terrestrial and aquatic systems in the eastern and upper Midwestern USA. The American Midland Naturalist. 2021;**186**(1):51-75

[19] Thompson LM, Troeh FR. Los suelos y su fertilidad. 4th ed. Barcelona, Spain: Reverté; 1982

[20] Goma-Tchimbakala J, Bernhard-Reversat F. Comparison of litter dynamics in three plantations of an indigenous timber-tree species (*Terminalia superba*) and a natural tropical forest in Mayombe, Congo. Forest Ecology and Management. 2006;**229**(1-3):304-313

[21] Freschet GT, Cornwell WK, Wardle DA, Elumeeva TG, Liu W, Jackson BG, et al. Linking litter decomposition of above-and below-ground organs to plant-soil feedbacks worldwide. Journal of Ecology. 2013;**101**(4):943-952

[22] Baraloto C, Timothy Paine C, Poorter L, Beauchene J, Bonal D, Domenach AM, et al. Decoupled leaf and stem economics in rain forest trees. Ecology Letters. 2010;**13**(11):1338-1347

[23] Jackson BG, Peltzer DA, Wardle DA. Are functional traits and litter decomposability coordinated across leaves, twigs and wood? A test using temperate rainforest tree species. Oikos. 2013;**122**(8):1131-1142

[24] Harmon ME. Woody detritus its contribution to carbon dynamics of old-growth forests: The temporal context. Old-Growth Forests. 2009;**207**:159-190

[25] Cornelissen JH, Sass-Klaassen U, Poorter L, van Geffen K, van Logtestijn RS, van Hal J, et al. Controls on coarse wood decay in temperate tree species: Birth of the LOGLIFE experiment. Ambio. 2012;**41**(3):231-245

[26] Erkan N, Comez A, Aydin AC, Denli O, Erkan S. Litterfall in relation to stand parameters and climatic factors in *Pinus brutia* forests in Turkey. Scandinavian Journal of Forest Research. 2018;**33**(4):338-346

[27] Pook E, Gill A, Moore P. Long-term variation of litter fall, canopy leaf area and flowering in a *Eucalyptus maculata* forest on the south coast of New South Wales. Australian Journal of Botany. 1997;**45**(5):737-755

[28] Zhang X, Wang W. Control of climate and litter quality on leaf litter decomposition in different climatic zones. Journal of Plant Research. 2015;**128**(5):791-802

[29] Wang H, Liu S, Mo J. Correlation between leaf litter and fine root decomposition among subtropical tree species. Plant and Soil. 2010;**335**(1):289-298

[30] Luo D, Cheng R, Shi Z, Wang W. Decomposition of leaves and fine roots in three subtropical plantations in China affected by litter substrate quality and soil microbial community. Forests. 2017;**8**(11):412

[31] Silver WL, Miya RK. Global patterns in root decomposition: Comparisons of climate and litter quality effects. Oecologia. 2001;**129**(3):407-419

[32] García-Palacios P, Prieto I, Ourcival J-M, Hättenschwiler S. Disentangling the litter quality and soil microbial contribution to leaf and fine root litter decomposition responses to reduced rainfall. Ecosystems. 2016;**19**(3):490-503

[33] Jiang L, Kou L, Li S. Alterations of early-stage decomposition of leaves and absorptive roots by deposition of nitrogen and phosphorus have contrasting mechanisms. Soil Biology and Biochemistry. 2018;**127**:213-222

[34] Sun T, Hobbie SE, Berg B, Zhang H, Wang Q, Wang Z, et al. Contrasting dynamics and trait controls in first-order root compared with leaf litter decomposition. Proceedings of the National Academy of Sciences. 2018;**115**(41):10392-10397

[35] Wang C, Chen Z, Ma C, Lin G, Han S. Three potential pathways influencing contrasting decomposition rates of fine roots. Journal of Beijing Forestry University. 2016;**38**(4):123-128

[36] He J, Wang Z, Fang J. Issues and prospects of belowground ecology with special reference to global climate change. Chinese Science Bulletin. 2004;**49**(18):1891-1899

[37] Leishman MR, Haslehurst T, Ares A, Baruch Z. Leaf trait relationships of native and invasive plants: Community- and global-scale comparisons. New Phytologist. 2007;**176**(3):635-643

[38] Funk JL, Nguyen MA, Standish RJ, Stock WD, Valladares F. Global resource acquisition patterns of invasive and native plant species do not hold at the regional scale in Mediterranean type ecosystems. Biological Invasions. 2017;**19**(4):1143-1151

[39] Pérez-Harguindeguy N, Díaz S, Cornelissen JH, Vendramini F, Cabido M, Castellanos A. Chemistry and toughness predict leaf litter decomposition rates over a wide spectrum of functional types and taxa in Central Argentina. Plant and Soil. 2000;**218**(1):21-30

[40] Hättenschwiler S, Jørgensen HB. Carbon quality rather than stoichiometry controls litter decomposition in a tropical rain forest. Journal of Ecology. 2010;**98**(4):754-763

[41] Gießelmann UC, Martins KG, Brändle M, Schädler M, Marques R, Brandl R. Lack of home-field advantage in the decomposition of leaf litter in the Atlantic rainforest of Brazil. Applied Soil Ecology. 2011;**49**:5-10

[42] Swift M, Heal O, Anderson J. Decomposition in Terrestrial Ccosystems. Oxford: Blackwell Scientific Publications; 1979

[43] Vitousek PM, Sanford RL. Nutrient cycling in moist tropical forest. Annual Review of Ecology and Systematics. 1986;**17**(1):137-167

[44] Wardle D, Lavelle P. Linkages between soil biota, plant litter quality and decomposition. In: Cadisch G, Giller KE, editors. Driven by Nature. Oxon, UK: CAB International; 1997. pp. 107-124

[45] Shaw MR, Harte J. Control of litter decomposition in a subalpine meadow-sagebrush steppe ecotone under climate change. Ecological Applications. 2001;**11**(4):1206-1223

[46] Blair J, Parmelee R, Beare M. Decay rates, nitrogen fluxes, and decomposer

communities of single-and mixed-species foliar litter. Ecology. 1990;**71**(5):1976-1985

[47] Smith V, Bradford M. Litter quality impacts on grassland litter decomposition are differently dependent on soil fauna across time. Applied Soil Ecology. 2003;**24**(2):197-203

[48] Taylor BR, Parkinson D, Parsons WF. Nitrogen and lignin content as predictors of litter decay rates: A microcosm test. Ecology. 1989;**70**(1):97-104

[49] Millar H, Smith F, Brown P. The rate of decomposition of various plant materials in soils. Journal of the American Society of Agronomy. 1936;**28**:914-923

[50] Minderman G. Addition, decomposition and accumulation of organic matter in forests. The Journal of Ecology. 1968;**56**:355-362

[51] Gartner TB, Cardon ZG. Decomposition dynamics in mixed-species leaf litter. Oikos. 2004;**104**(2):230-246

[52] Dilly O, Bloem J, Vos A, Munch JC. Bacterial diversity in agricultural soils during litter decomposition. Applied and Environmental Microbiology. 2004;**70**(1):468-474

[53] Kjoller A, Struwe S. Functional groups of microfungi in decomposition. In: Carroll GC, Wicklow DT, editors. The Fungal Community: its Organisation and Role in the Ecosystem. New York: Marcel Dekker; 1992. p. 619-630

[54] Dilly O, Bartsch S, Rosenbrock P, Buscot F, Munch JC. Shifts in physiological capabilities of the microbiota during the decomposition of leaf litter in a black alder (*Alnus glutinosa* (Gaertn.) L.) forest. Soil Biology and Biochemistry. 2001;**33**(7-8):921-930

[55] Harmon ME. Analysis of detritus and organic horizons for mineral and organic constituents. In: Robertson GP, Bledsoe CS, Coleman DC, editors. Standard Soil Methods for Long-Term Ecological Research. P S. New York: Oxford University Press; 1999. pp. 143-165

[56] Irmler U. Litterfall and nitrogen turnover in an Amazonian Blackwater inundation forest. Plant and Soil. 1982;**67**(1):355-358

[57] Giweta M. Role of litter production and its decomposition, and factors affecting the processes in a tropical forest ecosystem: A review. Journal of Ecology and Environment. 2020;**44**(1):1-9

[58] Pant S, Tiwari S. Litter fall and litter decomposition in a montane oak forest of Garhwal Himalaya. Tropical Ecology. 1992;**33**(1):103-109

[59] Sarjubala, Devi A, Yadava P. Wood and leaf litter decomposition of *Dipterocarpus tuberculatus* Roxb. in a tropical deciduous forest of Manipur, Northeast India. Current Science (00113891). 2007;**93**(2):243-246

[60] Tripathi O, Tripathi R. Litter production, decomposition and physico-chemical properties of soil in 3 developed agroforestry systems of Meghalaya, Northeast India. African Journal of Plant Science. 2009;**3**(8):160-167

[61] Kumar M, Joshi M, Todaria N. Regeneration status of a sub-tropical *Anogeissus latifolia* forest in Garhwal Himalaya, India. Journal of Forestry Research. 2010;**21**(4):439-444

[62] Meentemeyer V. Macroclimate and lignin control of litter decomposition rates. Ecology. 1978;**59**(3):465-472

[63] Gillon D, Joffre R, Ibrahima A. Initial litter properties and decay rate: A

microcosm experiment on Mediterranean species. Canadian Journal of Botany. 1994;**72**(7):946-954

[64] Joffre R, Ågren GI, Gillon D, Bosatta E. Organic matter quality in ecological studies: Theory meets experiment. Oikos. 2001;**93**(3):451-458

[65] Magid J, Cadisch G, Giller K. Short and medium term plant litter decomposition in a tropical Ultisol elucidated by physical fractionation in a dual 13C and 14C isotope study. Soil Biology and Biochemistry. 2002;**34**(9):1273-1281

[66] Helsen K, Smith SW, Brunet J, Cousins SA, De Frenne P, Kimberley A, et al. Impact of an invasive alien plant on litter decomposition along a latitudinal gradient. Ecosphere. 2018;**9**(1):e02097

[67] Ye X-Q, Yan Y-N, Wu M, Yu F-h. High capacity of nutrient accumulation by invasive *Solidago canadensis* in a coastal grassland. Frontiers in Plant Science. 2019;**10**:575

[68] Shad N, Liu Q, Fang H, Wang B, Gao Y, Liu X, et al. Soil sterilization and fertility impacts on urease and belowground mass specific phosphatase activity vary among Chinese tallow tree (*Triadica sebifera*) populations. Plant Ecology. 2022;**233**(4):397-406

[69] Melillo JM, Aber JD, Muratore JF. Nitrogen and lignin control of hardwood leaf litter decomposition dynamics. Ecology. 1982;**63**(3):621-626

[70] Patil M, Kumar A, Kumar P, Cheema NK, Kaur R, Bhatti R, et al. Comparative litter decomposability traits of selected native and exotic woody species from an urban environment of north-western Siwalik region, India. Scientific Reports. 2020;**10**(1):1-14

[71] Singh K. Studies in decomposition of leaf litter of important trees of tropical deciduous forests at Varanasi. Tropical Ecology. 1969;**10**:292-311

[72] Berg B, Staaf H. Leaching, accumulation and release of nitrogen in decomposing forest litter. Ecological Bulletins. 1981;**33**:163-178

[73] Vitousek PM, Turner DR, Parton WJ, Sanford RL. Litter decomposition on the Mauna Loa environmental matrix, Hawai'i: Patterns, mechanisms, and models. Ecology. 1994;**75**(2):418-429

[74] Liao C, Peng R, Luo Y, Zhou X, Wu X, Fang C, et al. Altered ecosystem carbon and nitrogen cycles by plant invasion: A meta-analysis. New Phytologist. 2008;**177**(3):706-714

[75] Broadbent AA, Orwin KH, Peltzer DA, Dickie IA, Mason NW, Ostle NJ, et al. Invasive N-fixer impacts on litter decomposition driven by changes to soil properties not litter quality. Ecosystems. 2017;**20**(6):1151-1163

[76] Suding KN, Lavorel S, Chapin Iii F, Cornelissen JH, Díaz S, Garnier E, et al. Scaling environmental change through the community-level: A trait-based response-and-effect framework for plants. Global Change Biology. 2008;**14**(5):1125-1140

[77] Cornelissen JH, Pérez-Harguindeguy N, Díaz S, Grime JP, Marzano B, Cabido M, et al. Leaf structure and defence control litter decomposition rate across species and life forms in regional floras on two continents. The New Phytologist. 1999;**143**(1):191-200

[78] Freschet GT, Aerts R, Cornelissen JH. A plant economics spectrum of litter decomposability. Functional Ecology. 2012;**26**(1):56-65

[79] Allison SD, Vitousek PM. Rapid nutrient cycling in leaf litter from

invasive plants in Hawai'i. Oecologia. 2004;**141**(4):612-619

[80] Zhang L, Wang H, Zou J, Rogers WE, Siemann E. Non-native plant litter enhances soil carbon dioxide emissions in an invaded annual grassland. PLoS One. 2014;**9**(3):e92301

[81] Chen Y, Sun J, Xie F, Yan Y, Wang X, Cheng G, et al. Non-additive effects of litter diversity on greenhouse gas emissions from alpine steppe soil in northern Tibet. Scientific Reports. 2015;**5**(1):1-11

[82] Zhang L, Xiaochi M, Hong W, Shuwei L, Siemann E, Jianwen Z. Soil respiration and litter decomposition increased following perennial forb invasion into an annual grassland. Pedosphere. 2016;**26**(4):567-576

[83] Zhang L, Zhang Y, Zou J, Siemann E. Decomposition of *Phragmites australis* litter retarded by invasive *Solidago canadensis* in mixtures: An antagonistic non-additive effect. Scientific Reports. 2014;**4**(1):1-8

[84] Pan J, Liu Y, Niu J, Fang H, Feng Y, Bai T, et al. Moso bamboo expansion reduced soil N_2O emissions while accelerated fine root litter decomposition: Contrasting non-additive effects. Plant and Soil. 2022:1-15. DOI: 10.1007/s11104-022-05785-8

[85] Chen B-M, Peng S-L, D'Antonio CM, Li D-J, Ren W-T. Non-additive effects on decomposition from mixing litter of the invasive *Mikania micrantha* HBK with native plants. PLoS One. 2013;**8**(6):e66289

[86] Zhang L, Wang S, Liu S, Liu X, Zou J, Siemann E. Perennial forb invasions alter greenhouse gas balance between ecosystem and atmosphere in an annual grassland in China. Science of the Total Environment. 2018;**642**:781-788

[87] Weidenhamer JD, Callaway RM. Direct and indirect effects of invasive plants on soil chemistry and ecosystem function. Journal of Chemical Ecology. 2010;**36**(1):59-69

[88] Shad N, Zhang L, Shah GM, Haifu F, Ilyas M, Ali A, et al. Plant invasion and N2O emission in Forest ecosystems. In: Advances in Forest Management under Global Change. London: IntechOpen; 2020

[89] Luan J, Liu S, Li S, Whalen JK, Wang Y, Wang J, et al. Functional diversity of decomposers modulates litter decomposition affected by plant invasion along a climate gradient. Journal of Ecology. 2021;**109**(3):1236-1249

[90] Yang R, Dong J, Li C, Wang L, Quan Q, Liu J. The decomposition process and nutrient release of invasive plant litter regulated by nutrient enrichment and water level change. PLoS One. 2021;**16**(5):e0250880

[91] Schmidt A, John K, Auge H, Brandl R, Horgan FG, Settele J, et al. Compensatory mechanisms of litter decomposition under alternating moisture regimes in tropical rice fields. Applied Soil Ecology. 2016;**107**:79-90

[92] Wallis E, Raulings E. Relationship between water regime and hummock-building by *Melaleuca ericifolia* and *Phragmites australis* in a brackish wetland. Aquatic Botany. 2011;**95**(3):182-188

[93] Song X, Peng C, Jiang H, Zhu Q, Wang W. Direct and indirect effects of UV-B exposure on litter decomposition: A meta-analysis. PLoS One. 2013;**8**(6):e68858

[94] Song X, Jiang H, Zhang Z, Zhou G, Zhang S, Peng C. Interactive effects of elevated UV-B radiation and N deposition on decomposition of Moso bamboo litter. Soil Biology and Biochemistry. 2014;**69**:11-16